CREATION AND METAPHYSICS
A GENETIC APPROACH TO EXISTENTIAL ACT

CREATION AND METAPHYSICS

A GENETIC APPROACH TO EXISTENTIAL ACT

by

HERVE J. THIBAULT, S.S.S.

Springer-Science+Business Media, B.V. 1970

Ursprünglich erschienen bei Martinus Nijhoff, The Hague, Netherlands 1970
Softcover reprint of the hardcover 1st edition 1970

ISBN 978-94-017-5816-1 ISBN 978-94-017-6255-7 (eBook)
DOI 10.1007/978-94-017-6255-7

To

C. T.

He who considers things genetically and originatively. . .
will obtain the clearest view of them.

(Aristotle, *Politics*, I, 2, 1252a24)

TABLE OF CONTENTS

PREFACE

During the last twenty-five years or so, studies in Thomistic existentialism have repeatedly indicated that the notion of creation played a decisive role in St. Thomas Aquinas' view of existence as an existential act or *actus essendi*. The importance for metaphysics of this view of existence as act warrants an investigation of the relation between creation and *actus essendi*; for St. Thomas is the only one, in the history of philosophy, to have considered existence as an act-of-being. This study will be limited to the early works of St. Thomas. By the time of the *Summa Contra Gentiles,* he had reached the key positions of his metaphysics. And the first fifty-three chapters of the *Summa Contra Gentiles* were written in Paris before June, 1259; the rest was completed in Italy before 1265.[1] The project was therefore conceived by St. Thomas during the first period of his career. How the notion of creation enabled him to transform the Aristotelian metaphysics of essence into a metaphysics of *esse* can be seen from three sections of the *Summa Contra Gentiles.*

Although primarily a theological treatise, the *Contra Gentiles* nevertheless accomplishes a radical metaphysical transformation of Aristotelianism by shifting the whole perspective from *esse in actu per formam* to *actus essendi.* Seen from the perspective of existential act as the absolute perfection, metaphysics is raised to a strictly transcendental plane of consideration. Admittedly, the *Contra Gentiles* was not written primarily to effect this change, yet the change pervades the whole work. In Book I, chapters 13–22, St. Thomas methodically passes from the consideration of God as Prime Mover to the "sublime truth" that He is Self-subsisting *Esse* or Pure Act of existence. In Book II, chapters 15–22, he shows that God alone can create, correcting an earlier opinion which he had defended in the *Commentary on the Sentences* concerning the possibility of angels acting instrumentally in

[1] Cf. A. Gauthier, *Saint Thomas d'Aquin, Contra Gentiles* (Paris: P. Lethielleux, 1961), Vol. I, pp. 33–34, 59; A. Walz, "L'Aquinate à Orvieto," in *Angelicum,* XXXV (1958), 181.

creation.[2] In Book III, chapters 64–76, the main proof of God's all-encompassing knowledge and providence is the totality of His causality. None of these three points is found in Aristotle.[3] Creation makes all the difference between the Judeo-Christian and the Greek world views.

The effect of this transformation, as far as an existential understanding of being is concerned, can be seen in the complete change it introduced in St. Thomas' theory of the analogy of being. In the *De Veritate*, 2, 11, he admitted only an analogy of proportionality between God and creatures through fear that analogy of proportion would compromise the infinite distance between the Creator and the creature and so tend to a univocal view of being: *finiti ad infinitum nulla est proportio*. By the time of the *In Boeth. de Trin.*, 1, 2, c, he had reached the conclusion that analogy is based on degrees of participation, *secundum magis et minus*, involving creative causality. Hence, he reversed his position: *est proportio creaturae ad Deum ut causati ad causam*. By itself, proportionality is insufficient. Were proportionality the key to the understanding of being, we should be left with an unexplained pluralism: with resemblances which are not accounted for. It is creation which binds being.[4] Proportionality is only a starting point, disclosing parallel essence/existence relationships among predicamental beings; but its explanation is found in the causal resemblance of creatures to their Creator from whom they hold their *esse*. Everything that exists, exists by virtue of an existential act or *actus essendi* which it holds from the Creator who is subsisting *esse*.[5]

There is no certainty as to when, where, or how St. Thomas reached this

[2] In II *Sent.*, I, 1, 3 (ed. Mandonnet, II, p. 53), and In IV *Sent.*, V, 1, 3 (ed. Moos, p. 209, n. 56).

[3] Cf. A. Mansion, "Le Dieu d'Aristote et le Dieu des chrétiens," in *La philosophie et ses problèmes*, Recueil d'études offert à Régis Jolivet (Paris: Em. Vitte, 1960), pp. 21–44.

[4] Cf. G. Klubertanz, *St. Thomas Aquinas on Analogy* (Chicago: Loyola University Press, 1960), pp. 27–29, 94–100. B. Montagnes, *La doctrine de l'analogie de l'être d'après saint Thomas d'Aquin* (Louvain: Publications Universitaires, 1963), pp. 81–93.

[5] St. Thomas shows how the notion of "act" is extended to existence in the following steps: (1) the origin of the notion of act from operations and activity; (2) the extension of the notion of act from activity to the substantial form which is the principle and also the term of activity; (3) the transposition of the notion of act from form to existence. Cf. *In Meta.*, IX, 5 (ed. Spiazzi, nn. 1826–27): Actus enim est de primis simplicibus; unde definiri non potest. Sed per proportionem aliquorum duorum ad invicem, potest videri quid est actus. Ut si accipiamus proportionem aedificantis ad aedificabile, et vigilantis ad dormientem, et ejus qui videt ad eum qui habet oculos clausos cum habeat potentiam visivam ..., et similiter per separationem ejus quod est praeparatum ad illud quod non est praeparatum, sive quod est elaboratum ad id quod non est elaboratum. Sed quorumlibet sic differentium altera pars erit actus, et altera potentia. Et ita proportionaliter ex particularibus exemplis possumus venire

conclusion that existence is an existential act. The communion of faith and reason in the theologian that St. Thomas was, and above all his reticence about himself, make it difficult to trace the evolution of his thought on this point. Certainly, his reflections on the secret name of God revealed in Exodus, 3:14, as well as his readings in Boethius, Dionysius, the *Book on Causes*, Avicenna, and his early effort to correct the universal hylomorphism of Avicebron, all contributed to his theory of existence as existential act or *actus essendi*. The works of C. Fabro, A. Forest, E. Gilson, J. de Finance, A. Hayen, R. Henle, J. Bofill have shown that it was by way of convergence from many angles, theological, metaphysical, epistemological, that St. Thomas concluded that *esse* is absolute act. All agree, however, that creation played a decisive role. In the circumstances, the aim of this study is to seek, especially in the early works of St. Thomas, a link between creation and existential act.

Now it is a well-known fact that St. Thomas always defended the possibility of a philosophical demonstration of creation, and that he equally defended the impossibility of a rational proof of the temporal origin of the universe. It should be possible, then, to extract from the writings of St. Thomas a theory of creation which is neither biblical nor theological, since the biblical notion of creation includes inception in time, but purely metaphysical, and which bears directly on existential act.

This work is neither an historical nor a textual study of St. Thomas.[6] Our task is limited to presenting one approach – a genetic approach – to existential act. A genetic method starts with *esse commune* and not with *ens in communi*. The existence of the things which compose the universe is the datum, and a genetic method seeks to account for existence as such (Chapter I). The insufficiency of physical processes of transformation and transmutation to account for existence, precisely because they presuppose some potential already existing, points to the need of a transcendental cause of *esse* (Chapter II). The first part of the argument consists in passing from the

ad cognoscendum quid sit actus et potentia. *De Potentia*, 1, 1, c: Nomen actus primo fuit attributum operationi . . ., exinde fuit translatum ad formam, inquantum forma est principium operationis et finis. Ibid., 7, 2, ad 9: Quaelibet forma signata non intelligitur in actu nisi per hoc quod esse ponitur. Nam humanitas vel igneitas potest considerari ut in potentia materiae existens, vel ut in virtute agentis, aut in intellectu; sed per hoc quod habet esse, efficitur actu existens. Unde patet quod hoc quod dico esse est actualitas omnium actuum, et propter hoc est perfectio omnium perfectionum.

[6] C. Fabro, *Participation et causalité selon s. Thomas d'Aquin* (Louvain: Publications Universitaires, 1961) contains an extensive historical study; L. Sweeney, "Existence/Essence in Thomas Aquinas's Early Writings," in *Proceedings of the American Catholic Philosophical Association*, XXXVII (1963), 97–131, started a textual study of the various lines of argument advanced by St. Thomas to establish that existence is an act.

biblical notion of creation which is couched in terms of production of the whole entity of things out of nothing and which entails temporal inception, to a metaphysical concept of creation which abstracts from both eternity or time and which can best be described as the emanation of *esse* from a source of *esse* (Chapter III). By keeping existence in perspective, a genetic method brings out the primacy of *esse* over essence and thereby reveals its character of act (ἐνέργεια) and perfection (ἐντελέχεια) in relation to essence; for *esse* is the actuality of the creative cause which is participated through creation (Chapter IV). A last chapter will attempt to show how predicamental causality shares in the transcendental causality of *esse* in bringing new beings into existence (Chapter V).

I wish here to express my gratitude to Fr. Gerard Smith, S.J., of Marquette University, to Fr. John Dowling, S.S.S., at Blessed Sacrament Seminary and John Carroll University, Cleveland, and to Mr. Gerald Steiner of General Electric Laboratories, Nela Park, for helpful discussions of the issues involved in this project; and to Fr. Michael Murray, S.J., Prof. Francis Collingwood, Prof. Edward Simons, Prof. Lottie Kendzierski, at Marquette University, for many suggestions. I am also very grateful to Fr. Joseph Roy Nearon, S.S.S., for help with the bibliography, to Mr. Edward Thuning of Scripps-Howard Broadcasting Company, Cleveland, for xeroxing copies of the manuscript, and to Miss Ardalee Bowers for invaluable secretarial assistance. And I want to thank the publishers, mentioned in the footnotes, for permission to quote copyright material: George Allen and Unwin, Ltd., to quote from M. Fakhry, *Islamic Occasionalism and its Critique by Averroes and Aquinas*; Basil Blackwell, to quote from F. Hoyle, *The Nature of the Universe*; The Clarendon Press, to quote from W. D. Ross, *Aristotle's Metaphysics*; Marquette University Press, to quote from M. J. Adler, *St. Thomas and the Gentiles*, and from Charles O'Neil, ed., *An Etienne Gilson Tribute*; William Morrow and Co., to quote from L. Barnett, *The Universe and Dr. Einstein*, published by William Sloane Associates; Martinus Nijhoff, to quote from W. Carlo, *The Ultimate Reducibility of Essence to Existence in Existential Metaphysics*, and from R. Henle, *St. Thomas and Platonism*, and from H. Spiegelberg, *The Phenomenological Movement*; Presses Universitaires, Louvain, to quote from S. Mansion, "Positions maîtresses d'Aristote," *Aristote et saint Thomas d'Aquin*; Sacred Heart Seminary Press, Detroit, to quote from A. Pegis, "St. Thomas and the Origin of the Idea of Creation," *Philosophy and the Modern Mind*, ed. Francis Canfield; Simon and Schuster, to quote from Bertrand Russell, *A History of Western Philosophy*.

Blessed Sacrament Seminary,
Cleveland, Ohio.

CHAPTER I

THE INVERSION OF METAPHYSICS

Is the notion of existence as an existential act, forming composition with the essence of individual beings, a conclusion capable of demonstration in metaphysics? Specifically, is there an organic link between a metaphysical notion of creation and existence as the act of existing? That is the problem, and the nature of the problem calls for a genetic method of investigation.

1. A Genetic Method

We find in the writings of St. Thomas two ways of approaching the problem of being: the noetic and the genetic. A noetic approach starts with the notion of being: the *conceptio entis*. It proceeds by analysis of the metaphysical structure of beings. The thought-movement is from essence to existence. Existence is not the focus of attention, but essence; although essence is viewed in relation to existence, actual or possible. The *De Ente et Essentia* is an example of a noetic approach. It starts with a logical analysis of material beings and gradually disengages their existence as neither contained within their essential notes nor flowing from the intrinsic principles of their nature. *Esse,* in other words, is neither part of the essence nor a property of the beings which we encounter in experience. And since existence does not belong to them by nature, yet they do exist, the only alternative is that they have existence from an extrinsic principle. Ultimately, this principle will needs be a subsisting act of existence: *quod est esse tantum.*[1] St. Thomas then concludes that *esse* is participated, as act in potency, in all finite beings whose essence is not their existence.

The argument thus moves from composition of essence and *esse* to participation of *esse,* and finally to subsisting or infinite *esse.*[2] The argument

[1] *De Ente et Essentia,* V (ed. Roland-Gosselin, p. 37:21 and p. 38:8).

[2] In chapter four, St. Thomas argued against the theory of hylomorphic composition in all creatures, including the angels, a theory which was widely held in his

presupposes that *esse* is an intrinsic act or actuating principle. This, of course, is the whole issue. Anybody knows the difference between *to be* and *not to be,* or between actual existence and mere possibility. But it is far from evident that actual existence is an *act of being*. As E. Gilson writes:

> It can be demonstrated that no essence is the cause of its own existence, from which it follows that whatever has an essence, and exists, must exist in virtue of an external cause; but no one has ever been able to demonstrate the conclusion that, in a caused substance, existence is a distinct element, other than essence, and its act.[3]

The problem for a noetic approach is to pass from actual existence to existence as the act of existing. For, in metaphysics, *esse* must be known as act and as effect prior to the demonstration that *esse* subsists as pure act of existing in a creative cause. In other words, metaphysics can conclude only *a posteriori* that the creator is subsisting *esse* or pure existential act. Let us then see whether a genetic approach can offer any help with this problem.

A genetic method deals with origins. In science, it traces patterns of historical development, such as the formation of the universe, the origin of life, the evolution of man. It deals, again, with the rise of nations, of cultures, of human institutions, etc. In metaphysics, a genetic method would seek to explain, not evolutionary processes ,but the very existence of things, i.e. being insofar as it *is*. The *Liber de Causis* is an example of this method. It is a Neo-Platonic account of being through creation. Its underlying theme is that every cause produces a proper effect, something unique and distinctive, so that the more universal is the effect, the more universal the cause. Thus being is the proper effect of the first Being who alone creates: *prima rerum creatarum est esse*.[4] All other agents, whether universal causes like the First

day and which he traced to the influence of the *Fons Vitae* of Avicebron. Cf. also *In II Sent.*, III, 1, 1, c (ed. Mandonnet, II, p. 86). Being, for Avicebron, is either simple or composite: if simple, it is God; if composite, it is made up of matter and form. The basic idea appealed to St. Thomas, but not its application. He transposed composition to the order of existence: all creatures are composed of essence and *esse*, God alone is pure *esse*. In the *De Ente et Essentia,* however, the argument is only dialectical. Cf. Ch. IV (ed. Roland-Gosselin, p. 34:15): Nisi forte sit aliqua res cujus quidditas sit ipsum suum esse . . .

[3] E. Gilson, *The Elements of Christian Philosophy* (New York: The New American Library, 1963), p. 140. There is no doubt that Gilson meant this remark as a challenge.

[4] *Esse* does not stand for existential act in the *Book on Causes*; its basic meaning is *ens* or essence, i.e. the most common aspect of all things, cf. L. Sweeney, "The Doctrine of Creation in *Liber de Causis,*" in *An Etienne Gilson Tribute* (Milwaukee: Marquette University Press, 1959), pp. 285–87; C. Fabro, *Participation et causalité* (Louvain: Publications Universitaires, 1961), pp. 236–39. When St. Thomas wrote his commentary on this work, he had discovered that it is a selection of thirty-two propositions from Proclus' *Elements of Theology* with glosses. And he was fully aware of its Platonic character: "According to the basic Platonic positions, which the author

Intelligence and the First Soul or particular agents on earth, presuppose being and produce in it other perfections or *bonitates* such as life, intellection, and formal determinations. Two orders of efficiency are thus clearly marked off: that of creation which gives being and that of secondary agents who introduce forms in pre-existing being. By the same token, the agency of the creative cause subtends that of all secondary causes, whether universal or particular.

Likewise, a genetic method distinguishes two orders of efficient causality: the physical and the transcendental, that of *fieri* and that of *esse*. The insufficiency of physical processes of transformation and transmutation to account for existence, precisely because they presuppose a potential or substrate already there, points to a transcendental cause of *esse*. A genetic method seeks beyond change and coming-to-be a source of existence as such. So doing – and this is the crux of the whole matter – a genetic method advances in the line of act. For change, according to Aristotle, is an act. It is the act of a potential tending to further actuation.[5] Change is not the addition of units, nor a series of intermediate stages, nor a succession of forms replacing one another; in fact, change does not belong to the order of essence or form, but to the order of existence. Change is an existential act, not fixed or abiding, but flowing and transient.[6] It is coming-to-be. That is why change is essentially in time, whereas forms are only indirectly in time.[7] Change is an imperfect existential act, *esse* is a terminal act. And St. Thomas adds that change can be understood by analogy as related to the subject that changes in the way that *esse* is related to the existent.[8] Both are existential acts.

But that is only half the story. By moving from physical causality to

of this work follows in many instances . . ., the most common aspects of things are caused by distinct principles, whereas what is more particular is effected by inferior principles." The *Book on Causes* bears traces of a Christian-Syriac or Arabian correction of the polytheism of Proclus to monotheistic creationism, cf. H. D. Saffrey, *S. Thomae de Aquino super Librum de Causis expositio* (Fribourg: Imprimerie S.-Paul, 1954), pp. XXX–XXXI; L. Gardet and M. Anawati, *Introduction à la théologie musulmane* (Paris: J. Vrin, 1948), pp. 246–47. Proclus, Porphyry, and Iamblichus opposed the Judeo-Christian doctrine of creation in time and defended the eternal emanation of beings, by way of cognitional overflow, through a descending order of intermediaries, cf. E. R. Dodds, *Proclus: The Elements of Theology* (2nd. ed., Oxford: The Clarendon Press, 1963), p. 290; P. de Labriolle, *La réaction païenne, Etude sur la polémique antichrétienne du Ier au VIe siècle* (Paris: L'artisan du livre, 1934), p. 12 ff.

[5] *Phys.*, III, 1, 201 a 10 – 202 a 12.

[6] Cf. P. Hoenen, *Cosmologia* (Romae: Apud Aedes Pont. Universitatis Gregorianae, ed. 5a, 1956), pp. 234–35, 573–76.

[7] *Phys.*, IV, 12, 221 b 20.

[8] In I *Sent.*, 19, 2, 2 (ed. Mandonnet, I, p. 470): Sicut motus est actus ipsius mobilis inquantum mobile est; ita esse est actus existentis inquantum ens est.

transcendental causality, a genetic method accomplishes a total inversion of the essence/existence principles of reality. Viewed from the angle of change and motion, *esse* appears as the ultimate and terminal act

> in which generation and all movement terminates, since every form and act is in potency before acquiring *esse*.[9]

But viewed from the angle of creation or absolute origin, *esse* is a primary act. It is only in the order of change or process that the potential precedes the actual; in the order of creation, there is no substrate to receive *esse*, but concomitance of co-principles: *esse* and determining essence. For, in creating,

> God at the same time gives *esse* and that which receives *esse*, so that his action does not require any pre-existing material to work on.[10]

From the viewpoint of change, *esse* is terminal act; from the viewpoint of creation, *esse* is primary. This is the inversion of metaphysics – *le renversement de la métaphysique* – mentioned by authors like Hayen, Fabro, de Finance, Henle, and Isaac.[11] Once *esse* is given its rightful primacy, essence is seen as the determining and limiting factor in any finite existent of its *esse*. Reality is basically *esse*, limited in varying degrees of ontological perfection by essence. In its order, essence is a positive perfection; but all its reality derives from its *esse*. Essence is the principle of diversification of *esse* in finite existents. As such, essence is distinct from and forms composition with *esse* in all finite beings.

To summarize: (1) a genetic method never has to pass from essence to existence considered as existential act; (2) a genetic method proceeds along the plane of act; from change, which is an imperfect existential act, to *esse*, which is the perfect existential act; (3) the transition from the physical order of change to the transcendental order of *esse* hinges on the need of creation, beyond processes, to explain the existence of the universe.

2. *Creation in metaphysics*

Historically, any consideration of creation in western philosophy can be

[9] *Contra Gentiles*, II, 52: Ad quod generatio et omnis motus terminatur: omnis enim forma et actus est in potentia antequam esse acquirat.

[10] *De Potentia*, 3, 1, ad 17: Deus simul dans esse, producit id quod esse recipit; et sic non oportet quod agat ex aliquo praeexistenti. Also I, 45, 4, ad 2: Compositum sic dicitur creari, quod simul cum omnibus suis principiis in esse producitur.

[11] A. Hayen, *La communication de l'être d'après saint Thomas d'Aquin*, Vol. II, L'ordre philosophique de saint Thomas (Paris: Desclée de Brouwer, 1959), p. 76: "Would it not be preferable to abandon the potency/act approach, applied analogically to matter and form and to essence and existence, as the key to metaphysics, and to substitute instead the two orders of causality, creation and process? The key to

traced to the influence of biblical revelation. Jewish, Christian, and Arab thinkers all testify to this fact. Does not this fact preclude even the possibility of philosophical investigation? That radical questioning, so justly called the essence of metaphysics, seems to be excluded by faith in the doctrine of creation. Martin Heidegger, for instance, remarks that the believer may perhaps repeat the Question of Being, but he cannot honestly be said to search for an answer. He is serenely sure that there is an answer to the problem of Being, and that faith furnishes the answer. The believer cannot experience the anguish of really questioning: "Why is there Being rather than nothing? Why existents? why not – nothing?" That is the Question of Being. And since metaphysics raises the question, metaphysics must answer it. Any reliance on extrinsic help blocks the enterprise. Without radically renouncing his faith and accepting the consequences of such a step, the believer is barred from metaphysical search.[12]

It is not that Heidegger objects to faith; but he wants to keep faith and reason incommunicable. He rejects a Christian Philosophy as a contradiction in terms.[13] He refuses to admit that philosophy and revelation can share

metaphysics would be, on the one hand, the interior order of *esse*, of creation and total causality, of the immanence of the Creator in his creation; on the other hand, the physical order of forms, of change and particular causes, attesting to the inner sustaining presence of God to their efficiency." C. Fabro, *Participation et causalité*, p. 290: "An uncritical application of the potency/act couplet to essence and *esse*, however handy and facile it may seem, would prove nothing and could lead to illusion. The real distinction between *esse* and essence stems from creationism. Now Aristotle, the theorist of potency and act, never considered *esse* or creation." J. de Finance, *Etre et agir* (Paris: Beauchesne, 1945), pp. 109–110: "[The originality of St. Thomas] does not consist in his having propounded a real distinction between essence and *esse*, since that distinction is already found in some Arab authors and in William of Auvergne. The originality of St. Thomas, as we see it, lies in his interpretation of that distinction in terms of (1) the theory of participation and (2) the theory of potency and act, but extended beyond its Aristotelian applications. More precisely, his originality lies in his view of *esse* as act." R. Henle, *Saint Thomas and Platonism* (The Hague: M. Nijhoff, 1956), p. 379: "When we turn to the Thomistic analogue to [Platonic] participation, namely, the relationship between creatures and God, so often designated in Saint Thomas by the same term, we find the explanation of the relationship to rest in the complex pattern of the four causes." J. Isaac, "L'unité de l'être," in *Bulletin thomiste*, VIII (1947-1953), pp. 52–55, reviewing E. Gilson's *L'être et l'essence*, remarked that any consideration of *esse* exclusively as ultimate act and complement of form cannot proceed beyond a dialectical stage. To rise to a truly transcendental plane of consideration, *esse* must be primary; for essence, in creatures, will then become the intrinsic measure of their *esse*.

[12] M. Heidegger, *Introduction to Metaphysics*, tr. Ralph Manheim (New York: Doubleday Anchor Books, 1961), p. 6.

[13] Ibid., pp. 6–7. Cf. H. Birault, "La foi et la pensée d'après Heidegger," in *Philosophies chrétiennes*, Recherches et débats, Cahier 10 (Paris: Arthème Fayard, 1955), pp. 108–132. Although urged with particular insistence by Heidegger, the

the same reality – the universe of human existence – and differ only according to the light in which each sees the same reality: the light of reason in the first instance, the light of faith in the second.

However, even when revelation gives answers to problems pertaining to the natural order, it does not supply any rational evidence for the answers. Revelation states facts; it offers no proof except the authority of the Speaker. So long as the evidence for creation is purely rational, so long as no appeal is made in the premises to biblical notions, the inquiry remains on the metaphysical plane.[14] It was in maintaining with the philosophers, against the common opinion of the theologians of his time, that the temporal origin of the universe cannot be demonstrated by reason, that St. Thomas was led to distinguish a metaphysical notion of creation from the biblical concept which includes temporal inception.[15] Heidegger has never given any consideration to this distinction, although it is an historical contribution to the Question of Being. He has persistently maintained that creation depends on faith not only for its origin but also for its meaning, and consequently that it finds no place in metaphysics.

However, the philosophical notion of creation is not derived from revelation. The biblical notion of creation is couched in terms of production *ex*

problem is not new. Bertrand Russell, *History of Western Philosophy* (New York: Simon and Schuster, 1945), p. 463, writes: "There is little of true philosophical spirit in Aquinas. He does not, like the Platonic Socrates, set out to follow wherever the argument may lead. He is not engaged in an inquiry, the result of which it is impossible to know in advance. Before he begins to philosophise, he already knows the truth: it is declared in the Catholic faith. If he can find apparently rational arguments for some parts of the faith, so much the better; if he cannot, he need only fall back on revelation. The finding of arguments for a conclusion given in advance is not philosophy, but special pleading." For the contrary opinion, cf. M. Adler, *St. Thomas and the Gentiles* (Milwaukee: Marquette University Press, 1938), p. 98: "It is precisely because of his Christian faith and his subjection to sacred theology as *scientia rectrix* that St. Thomas is a better metaphysician than Plato and Aristotle. To say that St. Thomas is primarily a theologian implies not a dogmatic solution of philosophical problems but rather a surer and more guarded consideration of essentially philosophic themes."

[14] Cf. L. Malevez, "Le philosophe et le croyant," in *Nouvelle Revue Théologique*, LXXXII (1960), 904; English tr. in *Philosophy Today*, V (1961), 14–30. "It is one thing to grasp and even to understand the simple natural meaning of a statement, and quite another thing to see the rational necessity or evidence for such a statement. Insight into the rational character of the statement: 'God created the universe,' is not given to reason through its faith in God the Creator. How could anyone doubt this? If, in the very act of being posited, faith in God the Creator gave reason as reason evidence that it is so and that God is necessarily the author of the universe, then faith would suppress itself. Reason would no longer have belief but evidence and science. Belief does not provide scientific evidence; and that is why the believer's reason, insofar as it is reason, remains unsated."

[15] Cf. In II *Sent.*, I, 1, 2, c (ed. Mandonnet, II, p. 18).

nihilo, and involves temporal origin. But if neither the eternity nor the temporal inception of the universe can be rationally demonstrated, as St. Thomas insisted, the metaphysical notion of creation must abstract from temporal origin or eternity, and retain only causal origin or dependence. St. Thomas sometimes used the Neo-Platonic term of emanation, *emanatio totius esse,* to describe creation. The term expresses metaphysical derivation and can be accommodated either to a temporal or an eternal universe. Moreover, emanation of *esse* implies a shift in meaning of creation. From an essentialist definition in terms of the production of things *ex nihilo,* one passes to an existential definition in terms of the flow of *esse* from a source of *esse.* Such a meaning is not the biblical concept of creation. It transposes the biblical notion into the existential order, and it leaves open the question of the first moment of creation. Why is this not philosophical?

The problems of disengaging a metaphysics of creation from the methods of approach and the problematics of sacred theology are serious. But St. Thomas can help us with the task. He carefully distinguished what can be known by faith alone from what can be demonstrated by reason; respecting both the autonomy of reason and the revelation of natural truths. Although never dissociated in St. Thomas' thought or writings, theology of faith and metaphysics remain specifically distinct. In the order of specification, metaphysics is metaphysics even when, in the order of exercise, it is practiced by a theologian.

There is, for instance, a long way from the revelation of God's secret name in Exodus, 3:14–15, to the sublime truth that God is *Ipsum Esse Subsistens* of the *Contra Gentiles,* I, 22. No Father of the Church, it seems, understood the text of Exodus as meaning that God is subsisting *Esse,* because no Father of the Church had a metaphysics of *actus essendi.*[16] Far from being

[16] Cf. Rouet de Journel, *Enchiridion Patristicum,* s.v. Exodus, 3:14, and *Index theologicus,* 94–95; X. Le Bachelet, "Dieu, sa nature d'après les Pères," in *Dictionnaire de théologie catholique,* IV, 1023–1154; F. Diekamp, *Die Gotteslehre in die katholische Dogmatik* (Freiburg: Herder, 1908), pp. 148–83. The Fathers of the Church refer to Exodus, 3:14, as a commonplace. But the Septuagint version had given the text an essentialist construction by translating: Σγώ εἰμι ὁ ὤν, which became τὸν ὄντα in the Book of Wisdom, 13:1, cf. A.-M. Dubarle, "La signification du nom de Iahweh," in *Revue des sciences philosophiques et théologiques,* XXXV (1951), 3–21; P. van Imschoot, *La théologie de l'ancien testament,* Tome I, Dieu (Tournai: Desclée de Brouwer, 1954), pp. 14–17. In general, the Fathers interpreted the divine name in terms of aseity (ἀγγεννητος), immutability, eternity. God is ὁ ὤν or τό ὀν or τὸ ὄντως ὄν, i.e. *ens realissimum.* The *Cohortatio ad Graecos,* 22, P.G. 6, 282, wrongly attributed to St. Justin, remarks that the expressions ὁ ὤν and τὸ ὄν are equivalent, and that Plato was amazed at realizing this on the occasion of his visit to Egypt. Plato, however, retained the neuter expression as less likely to shock his contemporaries. [On the origin of this legend, cf. P.G. 6, 282, n. 70 and P.L. 41, 236.] St. Justin, *Apologia,* I, 14, P.G. 6, 348: δ ὄντως θεὸς, i.e. the immutable, the unchanging. St. Clement of Alexandria, *Stromata,* V,

derived from Exodus, the notion of existential act is presupposed for that interpretation of the text. What St. Thomas marvelled at in the passage of the *Contra Gentiles* was that metaphysics could reach a conception of God which, in his mind, coincided with God's secret name.

What has to be established, and only metaphysics can do it, is that *esse* is an *act*: that *actus essendi* is not the mere fact of having been produced, *esse extra causas et extra nihilum*; nor is it a mere extrinsic denomination or a relation of dependence on a cause, nor a pure condition of essence, i.e. the state of reality as opposed to the state of possibility. What has to be established is the all-encompassing or intensive character of *esse* virtually-eminently containing all perfections and grounding all participations.[17]

6, P.G. 9, 60: ὁ ὢν καὶ ὁ ἐσομένος. St. Ephrem, *Adversus haereses*, 53, cf. Journel, n. 728: Hoc enim est nomen Dei, essentia quae non habet initium. St. Athanasius, *Epistula de decretis nycaenae synodi*, 11, P.G. 25, 441: ὁ δε θεὸς ὢν ἐστι κὰθ' ἑαυτόν, περιέχων τὰ πάντα. St. Epiphanius, *Adversus haereses*, 40, 5, P.G. 41, 685: Qui erat et est et semper existit, quemadmodum Moysi ipse declarat. The Cappadocians, opposing the Neo-Arian Eunomius for whom the only name of God was Pater Ingenitus, stress εγώ εἰμι ὁ ὤν: κατά τὸ ἄιδιον τε και ἄοριστον ἐν τὸ εἶναι. Theodoret, *In Exodum*, V, P.G. 80, 230: the eternally unchanging. St. John Damascene, *De Fide orthodoxa*, I, 9, P.G. 94, 836: πέλαγος οὐσίας ἄπειρον καὶ ἄοριστον, cf. Leo Sweeney, "John Damascene's Infinite Sea of Essence," in *Studia Patristica* (Berlin: Akademie Verlag, 1962), Vol. VI, pp. 248–63. St. Hilary, *De Trinitate*, I, 5, P.L. 10, 28: Ego sum qui sum.... Non enim aliud proprium magis Deo quam esse intelligitur; quia idipsum quod est neque desinentis est aliquando neque coepti. St. Jerome *In epistulam ad Ephesios*, 2, 3, 14, P.L. 26, 520: Qui est misit me ad vos.... Deus vero qui semper est, nec habet aliunde principium, et ipse sui origo est suaeque causa substantiae. St. Augustine, *Enarrationes in Psalmos*, 101, 2, 10, P.L. 37, 1311: Aeternitas ipsa Dei substantia est, quae nihil habet mutabile.... Non est ibi nisi, Est; non est ibi fuit vel erit..., sed quidquid ibi est, nonnisi est. Esset tibi nomen *ipsum esse*, nisi quidquid est aliud, tibi comparatum, inveniretur non esse vere. For St. Augustine, *Ipsum esse* means the immutable, cf. *De Trinitate*, I, 7, 5, P.L. 42, 942: Fortasse solum Deum dici oporteat essentiam. Est enim vere solus quia incommutabilis est, idque nomen famulo suo Moysi enuntiavit cum ait: Ego sum qui sum. Cf. R. Gagnebet, "Dieu, sujet de la théologie," in *Problemi scelti di theologia contemporanea* (Rome: Analecta Gregoriana, Vol. LXVIII, 1954), pp. 50–53. The influence of St. Augustine can be seen in St. Anselm, *Proslogion*, 2–22, P.L. 158, 227–238; for St. Anselm's insight into God's omnipresence in knowledge and in things, cf. A. C. Pegis, "The Argument of the Proslogion," in *Mediaeval Studies*, XXVIII (1966), 246. The Augustinian influence can be traced in St. Bernard, *De consideratione*, V, 6, 13, P.L. 182, 795; Richard of St. Victor, *De Trinitate*, I, 12, P.L. 196, 896; Peter Lombard, In I *Sent.*, 8, 1, (ed. Quaracchi, p. 57); Alexander of Hales, *Summa Theologica*, I, pars 2, inq. 3, tr. 1, qu. 1 (ed. Quaracchi, I, 517–18); St. Bonaventure, In I *Sent.*, II, Dub. IV (ed. Quaracchi, I, p. 60). The influence of Boethius can be seen in St. Albert, In I *Sent.*, II, 13, 2 (ed. A. Borgnet, XXV, p. 69): Esse essentiae, puro et simplici intellectu entis, Deo proprium est; cf. F. Catania, "Albert, Boethius, and Divine Infinity," n *Recherches de théologie ancienne et médiévale*, XXVIII (1961), 97–114.

[17] I-II, 2, 5, ad 2: Esse simpliciter acceptum, secundum quod includit in se omnem perfectionem essendi, praeeminet vitae et omnibus perfectionibus subsequentibus; sic igitur ipsum esse praehabet in se omnia bona subsequentia. – This all-inclusiveness of *esse* is variously described as *virtus essendi*: *Contra Gentiles*, II, 30; *In de div.*

To reject creation from metaphysics is to rule out such a view of *esse*. This is why Heidegger is not interested in that kind of investigation. For him, the problem of *esse* or *Sein* can be elucidated only by an analysis of the human situation, inasmuch as it is only in man that *being* becomes conscious and so unveils itself. This perspective causes Heidegger to shift consideration from the *Sein des Seiendes* to *Sinn von Sein*. The sense of being is its accessibility to man; it is the unveiling of being in human consciousness of existence.[18] And, for man, the sense of existence is ontological anxiety. For man is a being tending to death, projected into existence as doomed to die (*Sein zur Tode*). Being and non-being combine in man to make him radical contingency and, without a Creator, utter forlornness. But, one might ask, what about *Sein* or *esse* itself? Granted that human existence is ontological concern, what of *Sein itself* as the ultimate ground of all reality? Heidegger answers: temporality (Zeitlichkeit). All beings in our experience are limited and contingent. Nothing is permanent. Beings are projected into existence only to pass away. No matter where we turn, *Sein* reveals itself to us as temporal. The idea of an immutable and eternal being seems illegitimate.

J. B. Lotz has called attention to a shift of expression in the fifth edition of *What is Metaphysics?* Where the fourth edition read: "Being is independently of essents," the fifth edition has: "Being is never without essents." [19] The former reading exposed Heidegger to the charge of exaggerated realism, i.e. of reifying abstractions.[20] The later expression emphasizes the temporality and facticity of existence.

In Thomistic terms, Heidegger's phenomenological approach limits him to *esse commune* which is the *esse* of things which form the universe, in other words, the *esse* of creatures. Certainly, God cannot be bracketed with creatures under *esse commune* as Heidegger thinks that He must be – if He exists.[21] In Thomism, God is known from *esse commune* as the principle or

nom., 5, 1; *De Malo*, 16, 9, ad 5; as *perfectio essendi*: De Veritate, 29, 3; I, 4, 2, c.; as *plenitudo essendi*: In III *Sent.*, 13, 1, 2, 2, c.; *Contra Gentiles*, I, 43.

[18] Cf. L. Lynch, "Martin Heidegger: Language and Being" in *An Etienne Gilson Tribute*, p. 147: "Has the analysis presented, even in Heidegger's later writings, been of a *To-be* which is? Or has it been a descriptive analysis of man's awareness of *To-be* – of a *To-be* which is human experience? May the reason why metaphysics always seems to slip through his fingers be this: a *To-be* which *is* has never been touched? Is not the *To-be* of metaphysics hidden from the beginning? If it is, may it not be that existential analysis of this kind precludes the possibility of metaphysics instead of leading to the threshold of metaphysics?"

[19] J. B. Lotz, "Heidegger et l'être," in *Archives de Philosophie*, XIX (1955-1956, cahier 2), 14.

[20] The question has been raised, e.g. by H. Spiegelberg, *The Phenomenological Movement* (The Hague: M. Nijhoff, 1960), Vol. I, p. 287.

[21] Cf. *Brief über den Humanismus* (Bern: Francke, 1947), pp. 76, 86, 102–105.

cause of *esse*.[22] But Heidegger has eliminated creative causality from metaphysics. We are at the parting of the ways for a metaphysics of subjectivity and a metaphysics of creation.

If Heidegger has penetratingly disengaged the Question of Being – the *Sein des Seiendes* – his method nevertheless prevents him from going beyond *esse commune*. According to him, phenomenology, as the description of the human situation, and ontology, as the study of being, are the two aspects of metaphysics: phenomenology being the method and ontology the object. But his phenomenological analysis of *Da-sein* in terms of subjectivity makes him indifferent to any consideration of existence by way of conceptual objectisation. He probably would regard any description of existence as *actus essendi* as physicalism.

To answer the question, "Why is there being rather than nothing?" the method must be able to transcend the factual. In Heidegger, "mittence" or "thrownness" of beings in existence is factual. It is a datum, not an explanation. Some interpretation is needed. That is what a genetic method attempts to do.[23] It would be arbitrary to relegate all causal explanations to the sciences. Rather, one is led by the nature of the problem of existence to distinguish two orders of causality: the transcendental order of creation and conservation of beings in existence, and the predicamental order of transformations and transmutations of existing beings. The first is the metaphysical order; the second, the physical order. And, unless one resigns oneself to leave existence as an unexplained fact and as pure facticity, one has to look into the possibility of creation.

[22] *De Veritate,* 10, 11, ad 10: Ens quod est primum per communitatem, cum sit idem per essentiam rei cuilibet, nullius proportionem excedit; et ideo in cognitione cujuslibet rei ipsum cognoscitur. Sed ens quod est primum causalitate, excedit improportionaliter omnes alias res: unde per nullius alterius cognitionem sufficienter cognosci potest. *In de div. nom.,* 5, 2, n. 660: Omnia existentia continentur sub ipso esse communi, non autem Deus, sed magis esse commune continetur sub ejus virtute. *In Boeth. de Trin.,* 1, 2, ad 4: Deus autem quamvis non sit in genere intelligibilium, quasi sub genere comprehensum, utpote generis naturam participans, pertinet tamen ad hoc genus ut principium.

[23] Cf. K. Schmitz, "Metaphysical Restoration of Natural Things," in *An Etienne Gilson Tribute,* pp. 245–62: "To insist that the given meet the 'subjective conditions for the possibility of experience' is to take one's stand decisively in a subjective ground.... Efficient causality has no meaning in that context.... Causality has meaning only in a universe at whose center is a creative God and whose principiates are full-fledged beings. As long as human subjectivity remains the half understood *principium* of the metaphysical order, the entry of a transcendent God is excluded."

CHAPTER II

THE IRREDUCIBLE VALUE OF ESSE

"Why does something exist rather than nothing?" asked Leibniz. "For nothing is simpler and easier than something." [1] Leibniz, seeking the sufficient reason for beings, thought that he could discover it in the dependence of contingent beings on a necessary being: if there are contingent beings, then there must exist a necessary being. For Heidegger, the answer begs the question. It accounts for kinds of beings, but not for existence itself. No doubt, Heidegger has a point. So long as existence itself is merely assumed, no demonstration can establish more than necessity within its frame. In other words, granted that a situation involving contingent beings entails one or more necessary beings, the question remains: need such a situation exist at all? In the terminology of *Vom Wesen des Grundes*, the solution of Leibniz stays on the ontic level without rising to the truly ontological plane.

In the same way, Heidegger holds, the problem of causality arises on the ontic level, and, although this level presupposes the ontological, it cannot account for it. For that reason, Heidegger rejects any genetic method in metaphysics. In a critique of Aristotelian causality, he remarked that "a characteristic of the first presentation of the four causes is that the viewpoint does not make the basic distinction between transcendental grounding and specific ontic causes." [2] It is true that Aristotle has no theory of creation. His metaphysics of being *qua* substance cannot account for being insofar as it *is*. For when being is viewed from the angle of substantiality its principal act, precisely in that order, is form and not *esse*. Aristotle's explanations move from contingency to necessity, but always in an essentialist perspective.[3] On the other hand, creation elevates causality from the order of change and

[1] G. W. Leibniz, "The Principles of Nature and Grace according to Reason," *The Monadology and Other Philosophical Writings*, tr. Robert Latta (London: Oxford University Press, 1951), p. 415.

[2] *Vom Wesen des Grundes* (Halle: M. Niemeyer, 1929), p. 48.

[3] Cf. J. Chevalier, *La notion du nécessaire chez Aristote et chez ses prédécesseurs* (Paris: F. Alcan, 1915), pp. 135–36.

specific ontic causes to that of exisence. Creative causality is not affected by Heidegger's strictures. But this is where Heidegger demurs. At Cerisy-la-Salle, in 1955, M. de Gandillac, G. Marcel, P. Ricoeur confronted Heidegger with the possibility of properly philosophical issues arising historically from Judeo-Christian sources. But Heidegger evaded the question.[4]

In spite of that, the importance of Heidegger lies in the deliberateness and insistence with which he probes the problem of existence. He has repeatedly said that, for the first time in the history of western philosophy, the Question of Being has been raised *as a question*. The claim is understandable. For few realized, when Heidegger formulated his basic theme in *Sein und Zeit*, the primacy of *esse* in the metaphysics of St. Thomas. Even those scholastics who defended the real distinction between essence and existence in all finite beings, usually presented *esse* as a mode of essence. The merit and significance of Heidegger is to have, alone and independently, restored existence as the viewpoint of metaphysics. This perspective establishes metaphysics in its own right as a science of the real, and not of the abstract notion of being.

Heidegger's protest in recent years that he is not a metaphysician must be judged in the light of his criticism of traditional metaphysics as the forgetfulness of be-ing. His whole concern is with be-ing; and, although he seeks to clarify be-ing through human consciousness and subjectivity, be-ing is not thereby limited to man. All existents are "pro-jects" of *Sein*, so that *Sein* remains first. Man himself can grasp the meaning of his existence only through understanding being-in-itself. For man cannot understand himself as the one existent who raises the Question of Being unless he investigates what it means to be. Accordingly, anthropology, history, culture, language, all subserve the science of Being.

In order to emphasize the distinctiveness of be-ing, Heidegger has called it *Das Nichts*. But he has patiently explained that his use of the negative expression has primarily an historical import.[5] It is because essense or *whatness* has so completely dominated western philosophy that be-ing must be regarded as no-thing. *Das Nichts* is not pure nothingness, which Heidegger carefully calls *nihil negativum*, but refers to the peculiar way in which existence reveals itself to man. The piercing through of existence, in all its precariousness and non-essentiality, i.e. its negativity, is precisely the experience of the burning question: why being and not nothing? It is through its contingency and temporality that existence excites wonder and commands at-

[4] Cf. Th. Langan, "Heidegger in France," in *The Modern Schoolman*, XXXIII (1955-1956), 115–17.

[5] For a background study of this notion, cf. R. W. Mulligan, "A Note on Negativity," in *New Scholasticism*, XXXIII (1959), 162–83.

tention; it is through ontological *Angst* that man experiences his forlornness in this world. Because of philosophy's concentration on essence and neglect of existence, Heidegger resorts to a strange expression to shock us out of traditional ways of thinking. For him, existence reveals itself to human consciousness as burdened with finitude and contingency. Limiting himself to a phenomenological analysis, he refuses to speculate about any ultimate ground of existence. Rather, he seems to say: If you want positivism, then these are the facts.[6]

Many authors would deny any similarity between Heidegger's view of *Sein* and the Thomistic conception of *esse*. A. Dondeyne, for instance, warns that *esse*, whether conceived as *esse existentiae* according to the notional essence/existence perspective or as the *actus essendi* of St. Thomas himself, always refers to real existence in opposition both to intentional existence and to nothingness; but, for Heidegger, such a view of existence cannot be primary or original.[7] Granted; however, what Dondeyne overlooks is that *esse*, before it can be set over against intentional existence and non-being, is grasped in a living, existential judgement which is as immediate, underived, and irreducible as Heidegger's *un-veiling of Being*. The existential judgement, described by J. Maritain as a pre-metaphysical experience of existence, is as fundamental as *A-letheia*.[8] Evidently, *das Nichts* belongs to a different intellectual climate than Aquinas' *esse intensivum*, as does, for that matter, the Avicennian *esse accidentale*; yet, all these various characterizations of existence refer to one and the same objective reality: *to be*. Likewise, Heidegger's theme of the forgetfulness of be-ing, in spite of the dif-

[6] Useless to recall here the attacks of some logical positivists like R. Carnap, "Uberwindung der Metaphysik durch logische Analyse der Sprache," in *Erkenntnis*, II (1932), 219–41, which J. Maritain, *Quatre essais sur l'esprit dans sa condition charnelle* (Paris: Alsatia, 1956), p. 192, calls a brilliant piece of injustice; or A. J. Ayer, "The Genesis of Metaphysics," in *Philosophy and Analysis*, ed. M. Macdonald (New York: The Philosophical Library, 1954), pp. 23–25. More recently, P. Limke, J. Ebbinghaus, Ed. Springer, F. Heinemann have severely criticized Heidegger for what they dub his ontologism rooted in irrationalism, cf. L.-Ph. Ricard, "Le congrès de philosophie de Stuttgart," in *Archives de Philosophie*, XIX (Oct., 1955), 116–22. They charge Heidegger with relying more on language and fanciful etymologies than on the systematic elaboration of his presuppositions. Heidegger anticipated such charges in the postscript to *What is Metaphysics*? However, to say that the Question of Being would get scant attention, except for some disturbing personal experience, is easily verifiable.

[7] A. Dondeyne, "La différence ontologique chez M. Heidegger," in *Revue Philosophique de Louvain*, LVI (1958), 262–63.

[8] Cf. J. Maritain, *A Preface to Metaphysics* (New York: Sheed and Ward, 1948), p. 47; and *The Degrees of Knowledge* (New York: Charles Scribner's Sons, 1959), p. 279.

ference of approach, is none other than C. Fabro's *eclipse of esse.*[9] Both authors maintain that, historically, essence has stood in the way of *esse*; both are trying to bring *esse* to the fore of metaphysical consideration. What both authors are pointing to is remarkably well presented in the following passage from Coleridge's The Friend:

> Hast thou ever raised thy mind to the consideration of existence, in and by itself, as the mere act of existing? Hast thou ever said to thyself thoughtfully, It is! heedless in that moment, whether it were a man before thee, or a flower, or a grain of sand, – without reference, in short, to this or that particular mode or form of existence? If thou hast attained to this, thou wilt have felt the presence of a mystery, which must have fixed thy spirit in awe and wonder . . .[10]

It *is*! The experience reveals be-ing in all its empirical concreteness. *What* that something is, and how far and under what conditions it can be known, is the sphere of the sciences. But that something *is* – that constitutes an intrinsically metaphysical problem. The existential judgement is the experiental moment of metaphysics.

1. The existential judgement

It is only too easy to dismiss the Question of Being by taking existence for granted. After all, what we observe in the world around us are complex structures: men, flowers, or grains of sand. Nothing seems more natural than to reduce the complex to the simple, the structure to its units, the compound to the elementary; and, once the unit has been reached, to think that an understanding of the laws of combination will explain all that there is to know about the universe. But, just as the existence of the compound was taken for granted, so is the existence of the units presupposed. For Positivists, the universe is the given; existence is never questioned. One might say that science grants itself existence as a datum, whereas metaphysics deals directly with existence. If metaphysics is to be existential and not purely notional, it must deal with being as *what is* insofar as it *is*.

That is why an existential metaphysics starts from the judgement of existence. The existential judgement is altogether primary. True, is depends on direct sense perception and is in living continuity with sense experience; but, in its order, it is simple and underived. The existential judgement states and affirms existence: the fact or act (it does not matter which at this point) of

[9] C. Fabro, *Participation et causalité*, p. 280, says that he wrote his work to vindicate St. Thomas of *Vergessenheit des Seins*.

[10] S. T. Coleridge, "The Friend," in *The Complete Works* (New York: Harper, 1868), Vol. II, p. 463, quoted by H. Spiegelberg, *The Phenomenological Movement*, Vol. I, p. 285.

existing. Whereas simple apprehension abstracts from actual existence, the existential judgement grasps existence as exercised, *in actu exercito*, and that is why the existential judgement has to be assertive.[11]

In a lucid page in *Saint Thomas and Platonism*, R. Henle has shown the surplus intelligibility which the existential judgement adds to simple apprehension.[12] The problem is not to determine the meaning of the notion "existence" nor the use of the abstract term "existence" to designate a common property attributed to all existing things in virtue of the fact that they are; the real task is to show the import of the exercise of exisence by which whatever is, *is*. What lends weight to Henle's remarks is that they grow out of a long critique of the existential insufficiency of Platonism. The critique goes beyond the usual charge of reifying abstractions to show the radical impossibility of accounting for existence according to the *Via Platonica*. This shortcoming, common to all essentialisms, shows the importance of the existential judgement for metaphysics.

In our direct awareness of an existent, both simple apprehension and affirmation concur. The realization of existence, in apprehension, is virtually the affirmation of existence by the judgement. Later analysis may distinguish moments in the formation of the notion of being as *what-is*; but, in the intellect's grasp of an existent, there is involution of apprehension and affirmation, and both are in living continuity with the actual sense perception of a concrete existent. There is no juxtaposition, but compenetration of cognitive acts to form one total experience which finds expression in the existential judgement: It *is*.[13] The evidence is pre-categorial and pre-logical. It is the cognitive immediacy of the evidence of existence *in concreto*. Existence is an irreducible datum prior to any theorizing about its grounds.

[11] *In Peri Hermeneias*, I, lect. 5 (ed. Spiazzi, p. 29): . . . nam est, simpliciter dictum, significat in actu esse; et ideo significat per modum verbi.

[12] R. Henle, *Saint Thomas and Platonism*, p. 372.

[13] The importance of the existential judgement for a realistic metaphysics has been shown by R. Henle, *Method in Metaphysics* (Milwaukee: Marquette University Press, 1951), pp. 49–51, and "A Phenomenological Approach to Metaphysics," in *An Etienne Gilson Tribute*, pp. 68–85; E. Gilson, *L'être et l'essence* (Paris: J. Vrin, 1962), pp. 299–304; J. Maritain, *Existence and the Existent* (Garden City, N.Y.: Doubleday Image Books, 1957), pp. 35–37. No claim is here made that the existential judgement reveals existence as *actus essendi*, as though simple apprehension grasped an essence in abstraction, and then the judgement grasped its *actus essendi*. The existential judgement simply says: It *is*. It focuses on the existential dimension of the existent which was implicit in the apprehension and became explicit in the judgement. The explicitation of existence is the surplus intelligibility which the judgement adds to apprehension. These remarks seem to be an obvious answer to C. Fabro's questions in *Participation et causalité*, pp. 67–68, and "The Transcendentality of Ens-Esse and the Ground of Metaphysics," in *The International Philosophical Quarterly*, VI (1966), 424-26.

Any singular existent, insofar as it *is*, is perspicuous to metaphysical consideration, no matter how indefinite our grasp of its essential notes and characteristics may be. Reflection on each newly presented reality confirms the evidence that there is available to metaphysical consideration the common intelligibility of being insofar as it *is*. This common intelligibility is the *esse commune* of the things which surround us. An existential metaphysics requires in all its investigations continuity with its starting point: the concrete existential judgement.

Two remarks are in order at this point. (1) One concerns the first principles of knowledge. Every concrete existential judgement, such as: "It is" or "Things exist," contains the necessary intelligibility which can, on reflection, be formulated as the first principles of knowledge, namely, the principle of identity and the principle of non-contradiction. What needs to be emphasized is that these principles have not only an essentialist meaning, but primarily an existential import. Far from being mere tautologies or postulates, they express the law of being. The principle of identity not only asserts that "A = A," or that "every being is what it is," or that "a rose is a rose," but that "what is, *is*." And what is – insofar as it is – cannot not be. There is an intelligible necessity of being, present even in the most ephemeral existence, which is expressed in the principle of identity.[14]

The principle of non-contradiction, likewise, states that nothing can at once be and not be. It is between to be and not to be, absolutely speaking or existentially speaking, that there is no middle. Otherwise, there always seems to be a wedge between being and non-being: whether the μὴ ὄν which is and yet is not, or pure potency midways between nothingness and actuality, or self-constituting processes, or the indefinite of many-valued logics. It is only existentially understood that the principle of non-contradiction is exclusive. And when the principle of non-contradiction is taken existentially, one realizes the radicalism of Heidegger's Question of Being: Why are there essents rather than nothing?

The upshot of this digression is that existence is the evidence that grounds and controls every assertion or denial that we make. Existence is the light of being. As visibility depends at every moment on physical light, so intelligibility depends for every assertion on the evidence that what is, is, and what is not, is not. Any consistent system, from start to finish, depends on

[14] St. Thomas often remarked that nothing is so contingent that it does not involve some necessity. Cf. I, 19, 3, and I, 86, 3, for the famous examples of Socrates sitting or Socrates running. Socrates, although he need not sit or run, must, if he sits or runs, of necessity be sitting or running. Cf. P. Hoenen, "De connexionibus necessariis inter actus existentiales," in *Gregorianum*, XXXIV (1953), 603–39.

this light of being. Consistency, validity, truth depend – not only inceptionally or axiomatically, but at every step – on the evidence that what is, is. For that reason, the existential judgement is not trivial; it is the grounds of every statement which we make.

(2) A second point which shows the primacy of the existential judgement is the fact that it is presupposed by all attributive judgements which purport to say anything about existents. The truth, for instance, that Peter is a philosopher, a professor, a writer, etc., depends primarily on the truth that Peter is, and secondarily that he has those characteristics. The predicates of attributive judgements express various formal determinations of the unit which the existential judgement asserts *to be*. The predicates explicate our knowledge of the subject as existing under various forms. In that way, even the verb-copula of attributive judgements retains contact with concrete reality. And the mind, aware of its conformity with reality, knows that its judgements are true.

A verb-copula, devoid of existential reference, would link only abstract propositions, all having to do with mere possibles – as in a *mathesis universalis* or abstract geometry. For, of themselves, essences are intelligibles and not existents. The temptation of essentialisms is to reduce existential judgements to attributive judgements, as though to say that something *is* were to say that something belongs to the domain of real objects or that something is part of the universe. Such reductionism never comes to grips with the Question of Being. It presupposes existence. In other words, it begs the question.[15]

2. *The separation of esse*

The existential judgement, we have been saying, is the starting point of an

[15] Thus, in logical calculi, quantifiers, whether particular or universal, symbolize that a predicate applies in one case at least or in all cases. It is clear that this view presupposes existence. Ivo Thomas, "Existence and Coherence," in *Methodos,* II (1950), 76–84, has shown that, in the calculus of the *Principia Mathematica,* existence is purely formal, i.e. that it implies no more than coherence or absence of contradiction. A. Menne, "The Logical Analysis of Existence," in *Logico-Philosophical Studies* (Dordrecht: Reidel Publishing Company, 1962), pp. 88–96, explains how the formal concept of existence in logic cannot decide to which universe of discourse – real, historical, mathematical, mythical – an individual belongs. The real existence of trees in a clump, of chairs in a set, of stars in a galaxy is a matter of ontology. J. Dopp, "La notion d'existence dans la logique moderne," in *Proceedings of the Tenth International Philosophical Congress* (Amsterdam: North Holland Publishing Company, 1949), 735–39, has shown that it is possible, in accordance with Carnap's principle of tolerance, to enlarge logical symbolism so as to cover existential judgements. Cf. also, L. Apostel, "Logique et ontologie," in *Logique et analyse,* III (1960), 202-25.

existential metaphysics. The next step is to disengage noetically or to separate in knowledge *esse* from essence. This step is far from simple. It is only too easy to mistake metaphysical separation of *esse* for the abstraction of the general notion of what all things have in common, that is, the fact that they exist.[16] Separation is a technical term. It stands for a negative judgement; not any negative judgement, of course, but one which is the conclusion, reached after careful analysis of the things that form our empirical experience, that existence is not of the essence of finite and contingent beings.

The classical source of this doctrine of separation by means of a negative judgement is St. Thomas' *Expositio super librum Boethii de Trinitate,* V, 3. The doctrine is introduced to pinpoint the basic error of the Platonic theory of subsisting universals: the failure to distinguish what can be separated only in consideration and what can exist separately or independently of matter. In the next article, separation by negative judgement is used to free the subject matter of metaphysics – being and its properties, absolute perfections, potency and act – which need not, it is claimed, be bound up with matter or material conditions. The method of separation is not used in the *Expositio* to disengage *esse* from essence, but being-as-being from materiality. As a result, the separation of the *Expositio* can offer only a metaphysics of the notion of being: of *ens in communi.* In order to have a metaphysics of *esse commune,* one must start with existential judgements and practice separation on *esse* itself.

Usually, the separation of *esse* from essence is worked out, as in the first chapters of *De Ente et Essentia,* from a noetic approach. The method is heuristic. It leads to the insight that *what* a thing is cannot explain *that* it is. The insight is the conclusion that *esse* is not involved in quiddities because of quidditative necessity.[17] A genetic approach, on the other hand, starts

[16] Thus, for example, Morton White, "New Horizons in Philosophy," in *Saturday Evening Post,* CCXXXIII (Sept. 17, 1960), 24–25: "Physicists, some metaphysicians said, deal with material things, biologists with living things, mathematicians with numbers, points, and lines. But the metaphysician studies being as such. Not any particular being or limited class of beings, but just plain being.... It took philosophers a long time to realize that the number of interesting things that one can say about all things in one fell swoop is very limited. When you lump together such different items as kings, cabbages, bits of sealing wax, numbers, thoughts, and electrons in order to say what they all have in common, you are very likely to discover that what they have in common is the fact that they exist. You lapse into emptiness through the effort to become supremely general."

[17] For a detailed presentation of the separation of *esse* from essence, cf. Gerard Smith and Lottie Kendzierski, *The Philosophy of Being* (New York: The Macmillan Company, 1961), pp. 61–69, 81–87, 174–85, 254–63. The gist of the problem is given on p. 185. The difficulty of disengaging *esse* from its modes without prescinding from

with existence. It is through the realization that physical processes of change and evolution presuppose existence, but cannot account for it, that a genetic method disengages or separates *esse*. All processes – transformations and transmutations – presuppose something already there, i.e. a substrate. In transformations or transmutations, something passes from one state to another state, from one condition to another condition, from one form to another form; but the fact that processes presuppose something shows that they cannot account for existence. Even on the supposition that the universe has always existed, a supposition that cannot be rationally proved or disproved, the problem of existence cannot be dismissed by simply equating existence with the indestructibility of matter.[18] Indestructibles have only essential necessity. In other words, matter being matter or ultimate particles being ultimate particles, they are by nature indestructible – if they exist. But need they be? It is existential necessity that metaphysics looks for. The indestructibility of matter merely means that the amount of cosmic stuff will endure without loss throughout all the changes that form the processes of the universe. In addition, it may mean that, as far as reason can ascertain, the universe could be eternal. But the indestructibility of matter cannot explain the problem of being. As F. Hoyle writes:

> Consider the view of the out-and-out materialists. The appeal of their argument is based on simplicity. The Universe is here, they say, so let us take it for granted . . . The apparent simplicity, such as it is, of their case is only achieved by taking the existence of the Universe for granted. For myself there is a great deal more about the Universe that I should like to know. Why is the Universe as it is and not something else? Why is the Universe here at all? It is true that at present we have no clue to the answers to questions such as these, and it may be that the materialists are right in saying that no meaning can be attached to them. But throughout the history of science people have been asserting that such and such an issue is inherently beyond the scope of reasoned inquiry, and time after time they have been proved wrong.[19]

Hoyle's question: "Why is the Universe here at all?" is not essentially

any of them is explained on pp. 260–61. Disengaging *esse* does not end the metaphysical task, as Morton White seems to believe in the passage just quoted, but merely clears the way for the metaphysical enterprise.

[18] The issue is crucial for an existential metaphysics, especially for a metaphysics of *actus essendi*, as can be seen from this passage of R. W. Sellars, "Reformed Materialism and Intrinsic Endurance," in *The Philosophical Review*, LIII (1944), 377: "What the Neo-Thomist calls *esse* . . . is by the materialist assigned to dynamic and pattern-forming matter."

[19] F. Hoyle, *The Nature of the Universe* (Oxford: Blackwell, 1951), pp. 113–14. Hoyle has been criticized severely for this passage, cf. *Scientific American*, CXCV (Sept., 1956), 157–58. But Hoyle's motive in writing it cannot blur its message.

different from Heidegger's Question of Being, even though Hoyle would deny that it is a metaphysical question. Yet it is the question of existence. It confronts us with the same problem as we saw in the beginning of this chapter. It is the question of Leibniz. But whereas Leibniz searched the heavens for a necessary being, the materialist looks to matter. For the materialist, to be is to be a particle or a complex of particles. Complexes are contingent; particles are indestructible. But need such a situation exist at all? That is Heidegger's question, and it is also Hoyle's. Any attempt to explain existence by material causality is doomed from the start. As Lincoln Barnett observes:

> Even if one acquiesces to the idea of an immortal pulsating universe, within which the sun and earth and supergiant red stars are comparative newcomers, the problem of initial origin remains. It merely pushes the time of Creation into the infinite past. For while theorists have adduced mathematically impeccable accounts of the fabrication of galaxies, stars, star dust, atoms, and even the atom's components, every theory rests ultimately on the apriori assumption that something was already in existence – whether free neutrons, energy quanta, or simply blank inscrutable "world stuff," the cosmic essence, of which the multifarious universe was subsequently wrought.[20]

The interest of these remarks, so far as our problem is concerned, lies in the link indicated between existence and creation. To dismiss the question of existence is to reject the need of creation; and, vice versa, to reduce creation to physical processes is to miss the Question of Being. To seek the ground of existence is not to look for more fundamental material causes, which always presuppose existence, but for a source of existence.

From a genetic approach, the separation of *esse* is effected through the negative judgement that the existence of the universe is not accounted for by physical processes of change and transformation. Processes bring new beings into existence out of something already there. This fact forces one to seek a cause of existence as such; that is, a cause of existence without presupposition. For it is existential necessity that metaphysics seeks. The whole metaphysical project is a relentless search for existential necessity. The Atomists situated existential necessity in the indestructibility of indivisibles. Plato and Aristotle, in different ways, situated necessity in the self-identity of pure forms: the necessary is the immutable or everlastingly self-identical. It is the merit of Heidegger's critique of western philosophy to have shown the existential insufficiency of such world-views. Nevertheless, metaphysical reflec-

[20] Lincoln Barnett, *The Universe and Dr. Einstein* (New York: Mentor Books, 1954), p. 115.

tion on the intentional thrust – the *intentio profundior* [21] – of the systems of Plato and Aristotle suggests that unless *esse* subsist in itself (according to the intent of Platonism) or be pure existential act (according to the intent of Aristotelianism), we shall never find existential necessity.

3. From the Fact of Existence to Existential Act

Once the Question of Being has been raised and the primacy of existence has been focused, the next problem is: is existence merely a fact, a datum, a permanent possibility of sense experience, a presence, a being-there, or is existence an intrinsic perfection and the primary actuation of a being?

In a metaphysics of *esse commune*, existence is seen as the universal aspect of all things which is not explained by processes, since processes require something pre-existing. This insight leads to a further conclusion, namely, that no receptive potency precedes, even by nature, existence. Existence is altogether primary and originative. And if act is, absolutely speaking, prior to potency, then the primacy of existence would indicate that existence is an existential act.

We have already mentioned that Aristotle described change as the actuation of a potential. In terms of being, change involves potency and act. Change is a transitional act: the act of a being tending to further actuation. Change is more than an event, an occurrence, a happening; it is the progressive actuating of a subject.[22] This gradual actuation is an existential act. New beings come into existence through changes. But, because changes affect a subject, they are possible only within existence, they depend on existence, they bathe in existence. And if change is an intrinsic act in a subject, all the more so will existence be an intrinsic act. For change is an imperfect act; existence is absolute act. Motion and change keep the universe evolving only within a matrix of existence which maintains the universe in actuality.

[21] Aristotle speaks of the inner thrust of a system carrying the thought beyond the expression. In *Meta.*, A, 4, 985 a 4–5, in reference to Empedocles who had introduced love (attraction) and strife (repulsion) as explanatory principles of processes, he wrote: "If one takes what Empedocles says according to its *intent* rather than its obscure expression...." St. Thomas likewise speaks of the inner drive of St. Augustine's meaning beyond its systematic expression in *De Spir. Creat.*, 10, ad 8: "Ut profundius intentionem Augustini scrutemur, et quomodo se habeat veritas circa hoc, sciendum est...." The *intentio profundior* has become a famous theme with A. Hayen, *La communication de l'être,* pp. 45–122. For an appreciation of Hayen's method, cf. M. D. Chenu, *Bulletin thomiste,* X (1957-1959), n. 2353, pp. 957–58.

[22] Cf. *Physics,* III, 2, 201 b 32; *Meta.*, IX, 6, 1048 b 18–36. For instantaneous changes, cf. Ch. De Koninck, "Le devenir instantané," *Revue thomiste,* LXII (1962), 423–27.

As presupposed by evolutionary processes, existence is absolutely primary; as actuating whatever is in any way whatsoever, existence is all-pervading. *To be* is primal act and perfection. Consideration of change as act in the existential order leads to the conclusion that existence is an existential act.

Once the problem of actuality is situated squarely in the existential order, the notions of act and actuating become existentialized. They transcend the ontic order. *To be*, absolutely speaking, does not mean to be a pure form or a composite of form and matter; it means either to be a pure act of existing or, at least, to have an act of existing. This conclusion can, perhaps, be seen by contrast. What, for instance, does *to be* mean for Aristotle. Commenting on the various meanings of being listed in *Metaphysics*, D, 7, 1017 a 7–23, W. D. Ross presents Aristotle's answer:

> It may seem surprising that Aristotle, while dwelling on the two main senses of the copulative "is" – those in which it indicates respectively accidental and essential being – should say nothing of the existential "is," which nevertheless is presupposed in his account of accidental being (cf. 11. 19, 20, 21). The reaon is that, though logically the existential "is" may be distinguished from the copulative, metaphysically it is not. *To be* is either to be a substance, or to be a quality, or to be in some other of the categories, for nothing can be without being of some kind.[23]

What happens to the *to be* of the existential judgements in such a view of existence? It is taken for granted by being absorbed in the actuality of the essence. But such a viewpoint remains on the ontic level. Basically, it is the same kind of reductionism as the view which claims that *to be* is to be a particle or a complex of particles. Such explanations bypass the question of existence only by presupposing it, i.e. by taking existence for granted. Existence is irreducible to essence or matter. Ontologically, it can be described only as act in the existential order or, better still, as existential act.

What confirms that conclusion, from a genetic viewpoint, is the existential character and act-quality of causality. In Aristotelianism, the dynamism of nature is due to act. An agent operates only insofar as it is in act, and not in potency. The act, by which the agent operates, determines the kind of effect that it produces. Like causes, like effects.[24] Formal acts account for transformations and transmutations in the universe. But predicamental causality requires existence. What about the existence of the things which make up the

[23] W. D. Ross, *Aristotle's Metaphysics* (Oxford: The Clarendon Press, 1958), Vol. I, p. 308.

[24] The axiom, like begets like, is shown to be grounded in *act* in this passage from the *Contra Gentiles*, I, 29: De natura actionis est ut agens sibi simile agat, cum unumquodque agat secundum quod actu est.

universe? Existence everywhere appears limited, particular, individuated. And if existence is irreducible to essence or to matter, it requires a proper or *per se* cause. That cause will needs be, not simply a cause of things, but precisely a cause of the existence of the universe. That cause will have to be entirely existential, because of the primacy of its proper effect, and it will have to be wholly act, because of the universality of its proper effect. That cause, as we shall see in Chapter 4, will have to be subsisting *Esse*. Existence thus discloses itself as pure act in its source. And the primacy of existence distinguishes creation from all other modes of production. By relating creation directly and immediately to existence, a genetic method lifts causality out of the ontic order and situates it in the transcendental order of existence.

The net result is that *esse* can now be regarded as basic reality – the primary ontological reality – which all finite beings share in various measures. The whole realm of finite beings participates, through created *esse*, in the absolute perfection of subsisting *Esse*. All things are in total and constant dependence for their existence on the creative source of *esse*. A creature is always actually participating in *esse*. It never breaks loose from its source as a ripe fruit from a tree. Existents are never self-contained and self-sufficient: their *esse* is theirs only in the sense that it is in them the actuating principle of their being, but never in the sense that it is theirs independently and autonomously.

In that existential perspective, essence is less what receives or has *esse* as what, in *esse*, limits and diversifies *esse*. Essence has no reality or existential status apart from *esse*. *Esse* is primary act; and essence, in a finite existent [*ens*], is the intrinsic limiting principle of its *esse*. For, any determination of *esse* is a limitation of *esse*.[25]

[25] The relation of *esse* and essence is very difficult to express. G. Phelan, "The Being of Creatures," in *Proceedings of the American Catholic Philosophical Association*, XXXI (1957), 121, warns against the dangers of category mistakes. Since the terms which we use to designate metaphysical principles, such as act and potency, existence and essence, are grammatical nouns and substantives, it is difficult not to think that what they signify are things. Some authors tend to regard essence as a privative limit of *esse*, e.g. W. Carlo, *The Ultimate Reducibility of Essence to Existence in Existential Metaphysics* (The Hague: M. Nijhoff, 1966), pp. 103–104: "This is what we mean when we say that essence is the intrinsic limitation of *esse*. It is not *that which* limits *esse*, it is the *limitation* of *esse*; it is not *that which* receives, determines and specifies *esse*, it is the very specification of existence." Others stress the positive character of essence in *esse*; cf. M. Corvez, "Existence et essence," in *Revue thomiste*, LI (1951), 322–23: "Existence does not absorb essence, nor are they identified in reality. Essence always remains the potential which existence actuates. True, essence is not in potency to receive *esse*; nevertheless, in an existent, essence is the potential principle which forms real composition with *esse*. Potency in actual existence does not cease to be the actuated; it remains a real capacity limiting act. It is the same with prime matter informed by a substantial form, or any receptive capacity in union

One last remark. St. Thomas always refused to identify *esse* with the mere fact of having been caused or with the causal relation of a being to its source.[26] Actuality is intrinsic to things. Whether form or existential act, the actuating principle is immanent to the subject that is thereby actualized. Thus the position of St. Thomas is not affected by Heidegger's criticism of medieval metaphysics. Heidegger claims that *esse*, for the scholastics, meant being-in-act (*esse actu*) as opposed to potential being (*esse potentia*). Actuality then becomes *positio extra causas*, and Heidegger sees in this fact the possibility, for medieval man, of correlating *esse actu* with creation. It is in virtue of his faith, and not of metaphysics, that a believer expresses in Aristotelian terms the doctrine of creation. But this, for Heidegger, is not a legitimate extension of Aristotelian ἐνέργεια to cover existence, but a theological interpretation of Aristotelianism in the light of biblical creation. In confirmation of his criticism, Heidegger can point to Suarez who conceived existence as the state of actuality due to production.[27] The example is well chosen. The essentialism of Suarez is historical.[28] Heidegger could have added Henry of Ghent and Duns Scotus, but not St. Thomas. For St. Thomas, creation opened up the existential view of being which, by fusion of the doctrines of act and participation, resulted in his theory of existential act or *actus essendi*.

with the principle that actualizes it. The real distinction between essence and *esse* need not imply that essence and *esse* can be actually separated, unless one blandly refuses to admit any real distinction except between separable elements." The position adopted in this study is that essence is not purely a privative limit of *esse*, but a positive intrinsic principle, in an existent, determining its *esse*.

[26] The metaphysical insufficiency of such a position can be seen in *Contra Gentiles*, II, 52: "To be caused by another does not appertain to being-as-such; otherwise, every being would have to be caused by another, so that we would have to proceed to infinity in causes. . . . Therefore, the *esse* which subsists must be uncaused, and hence no caused being is its *esse*. C.f *De Pot*., 3, 5, ad 1; 1, 44, 1, ad 1.

[27] Cf. M. Heidegger, *Nietzsche* (Pfullingen: Neske, 1961), Vol. II, pp. 418–19.

[28] For the position of Suarez, cf. *Disputationes Metaphysicae*, Disp. XXXI, sect. 6, n. 23 (ed. Vives, XXVI, p. 250): Quia essentia creaturae non hoc necessario habet ex vi sua ut sit actualis entitas, ideo quando recipit entitatem suam, concipimus aliquid esse in ipsa quod sit illi formalis ratio essendi extra causas; et illud sub tali ratione appelamus existentiam, quod licet in re non sit aliud ab ipsamet entitate essentiae, sub diversa tamen ratione et descriptione a nobis concipitur, quod ad distinctionem rationis sufficit.

CHAPTER III

TWO VIEWS OF CREATION

In philosophy, the view of existence as something accidental to the nature of things and accruing to them from a creative source is due to Alfarabi and Avicenna. Alfarabi was the first philosopher to transpose into the metaphysical order the Aristotelian distinction between knowledge of what a thing is and knowledge of its existence. He interpreted this logical distinction as an indication of a real difference in things between their essence and their existence. In this distinction, he saw what separates the Necessary Being from all contingent beings and the dependence of all finite beings on God. Alfarabi did not explain how he regarded existence, except to say that it is accidental in contingent beings. Nevertheless, if he did not exploit the implications of his distinction, he added a new dimension to metaphysics and started an age-long controversy.[1] Its first echo is heard in the opposition of Averroes to Algazali and Avicenna.

1. Avicenna

Avicenna refined and elaborated in his own personal way the existential insight of Alfarabi. The core of Avicenna's metaphysics is the theory of the

[1] I. Madkour, *La place d'Alfarabi dans l'école philosophique musulmane* (Paris: Adrien-Maisonneuve, 1934), p. 221; J. Chevallier, *Histoire de la pensée* (Paris: Flammarion, 1960), Vol. II, pp. 262–63; E. Gilson, *A History of Christian Philosophy in the Middle Ages* (New York: Random House, 1955), p. 185. It may be that the *Gem of Wisdom* is not Alfarabi's but Avicenna's; however, the same doctrine in found in the *Fontes Questionum* recognized as a work of Alfarabi, cf. M. Cruz Hernandes, "El Fontes Questionum," in *Archives d'histoire doctrinale et littéraire du moyen âge,* XVIII (1951), 304–5, with the Latin text on pp. 316–18; A. Goichon, *La distinction de l'essence et de l'existence d'après Ibn Sina* (Paris: Desclée de Brouwer, 1937), pp. 151–55. On the discussed authorship of the *Gem of Wisdom,* cf. M. T. D'Alverny, "Anniya-Anitas," in *Mélanges offerts à E. Gilson* (Paris: J. Vrin, 1959), 71–72; for references, N. Rescher, *Al-Farabi, An Annotated Bibliography* (Pittsburgh: The University of Pittsburgh Press, 1962), s.v. Strauss (1934), Georr (1941-46), Pines (1951).

possibles. Of themselves, possibles are midway between necessity and impossibility. The necessary cannot not be; the possible can either be or not be; the impossible cannot be. All three divisions relate to existence. Essences, by themselves, are possibles. Although they have no inherent claim to existence, essences have an inner consistency of their own: their possibility belongs to them as such; they are not nothing. Existence, however, must come to them from some extrinsic principle, and ultimately from *Ipsum Necesse Esse*.

> In order to exist, a thing must be intrinsically possible; otherwise, it could never be. No agent could produce it unless it were possible in itself . . . Hence, its possibility-to-be is its own and cannot be due to any extrinsic principle; but its necessary existence it holds from its cause.[2]

From this analysis, Avicenna draws two conclusions which bear on our problem. The first is that in every being, other than the Necessary Being, existence is an accident of its essence. All existence emanates by way of cognitional overflow from a unique Necessary Being, through ten intermediary Intelligences, down to the last which is the Giver of Forms. All are creators and the process is eternal and necessary. The second conclusion of Avicenna is the clear distinction between two orders of causality: that of agent cause and that of moving cause. The first is creative; the other, physical. The first gives existence; the second effects transformations. Although Avicenna uses all the common Arab terms for creation, his inspiration remains Neo-Platonic. It owes nothing to "the metaphysics of Exodus." His system does not rest on any conception of a Creator who is *Ipsum Esse*, but who is the One of the pseudo *Theology of Aristotle* and so of the *Enneads* of Plotinus. Avicenna's problem is to explain how the many can proceed from the One. His answer is the theory of mediating emanations as in the *Theology of Aristotle*.[3]

[2] Avicenna, *Metaphysica* (Opera per canonicos regulares s. Augustini emendata, Venitiis, 1508), IV, 2, fol. 85v: Omne enim quod incipit esse, antequam fit, necesse est ut sit possibile in se. Si enim non fuerit possible in se, illud non erit ullo modo. Non est autem possibilitas sui esse eo quod agens sit potens supra illud cum ipsum non fuerit possible. *Metaphysica*, IX, 4, fol. 104v: Possibilitas sui esse est ei propter se, non propter principium, sed est ei a principio necessitas sui esse. The metaphysical essentialism of Avicenna has often been scored: L. Gardet, "En l'honneur du millénaire d'Avicenne," *Revue thomiste*, LI (1951), 335–37; E. Gilson, *Being and Some Philosophers* (Toronto: Pontifical Institute of Mediaeval Studies, 1949), pp. 75–80; G. Smith, "Avicenna and the Possibles," in *New Scholasticism*, XVII (1943), 346–48; B. Zedler, "St. Thomas and Avicenna in the *De Potentia Dei*," in *Traditio*, VI (1948), 152–56.

[3] G. Vajda, "Notes d'Avicenne sur la Théologie d'Aristote," in *Revue thomiste*, LI (1951), 346–406, and comments of L. Gardet, pp. 335–40; on Avicenna's theory of

Given these two points – namely, the distinction of essence and existence and the further distinction of two orders of causality – Avicenna was able to formulate his theory of the sructure of being in terms of his theory of causality. Avicenna credited the Kalam theologians with the view of the radical contigency and nothingness of creatures by themselves. But, whereas the theologians concluded to creation in time, since all creatures must pass from nothingness to be-ing, Avicenna was forced by his theory of cognitional emanation to maintain an eternal and necessary process of creation. For every possible must emanate in existence just as it is necessarily known by the Supreme Being. Accordingly, Avicenna transposed the theologian's consideration of the creature's passing from non-being to existence from the temporal order to the existential structure of things in themselves. The priority became one of nature and not of time. Creation does not involve temporal succession from non-being to be-ing, but composition, in nature, between an essence which of itself would never be and its existence which comes from an extrinsic principle. This transposition was facilitated by the fact that Avicenna understood potency in a Neo-Platonic way as privation, lack of being, and defect. He did not consider essence as the potential, in existents, limiting their *esse*, but as a possible which, of itself, tends to nothingness unless posited in existence by its creative cause:

> Essences do not deserve to be; considered in themselves and apart from their relation to the Necessary Being, they deserve privation of being, and so, in themselves, they are all false. It is to Him alone that they owe their reality [certitudo].[4]

Avicenna accordingly regarded *esse* as a *lazim 'amm* or transcendental concomitant of essences due to an extrinsic principle. As such, *esse* must be considered as other than essence.[5] Avicenna's rigid existential determinism is thus counterbalanced by the real distinction of *esse* and essence which stamps all creatures and sets them apart from the Necessary Being. The

creative emanations, D. Saliba, *Etude sur la métaphysique d'Avicenne* (Paris: Presses Universitaires, 1926), pp. 125–46; A. Goichon, *La distinction de l'essence et de l'existence d'après Ibn Sina,* pp. 224–43; L. Gardet, "Quelques aspects de la pensée avicennienne," in *Revue thomiste,* XLVII (1939), 552–63. There is no evidence that Avicenna knew of the *Book on Causes*; but he knew Proclus, cf. L. Gardet et M. Anawati, *Introduction à la théologie musulmane,* (Paris: J. Vrin, 1948), pp. 246–47.

[4] *Metaphysica,* VIII, 6, fol. 100v.

[5] A. Goichon, *La distinction de l'essence et de l'existence d'après Ibn Sina,* pp. 114–23; M. D. Roland-Gosselin, *Le "De Ente et Essentia" de s. Thomas d'Aquin* (Paris: J. Vrin, 1948), pp. 151–56; A. Forest, *La structure métaphysique du concret selon saint Thomas d'Aquin* (Paris: J. Vrin, 1956), pp. 140–41, 160–65.

Necessary Being has no essence; creatures have an essence and so possibility to be or not to be.[6]

J. Paulus has questioned whether Avicenna did claim more than a logical distinction between *esse* and essence. According to Paulus, epistemology in Avicenna so dominates and completely subordinates ontology, that his perspective is the intelligible and not the existential.[7] While it is true that Avicenna did take an ultra realistic view of essences, he had no choice in the matter. Since his emanative determinism implied necessity *ab alio* of all possibles *a se*, Avicenna's only way of admitting contingency in creatures was to place it in essences. Their inalienable self-identity, as possibles, precludes pantheism in a strict sense. This radical essentialism is what prevented Avicenna from seeing existence as *actus essendi*. What is questionable is this: unless *esse* is viewed as act, what meaning can a real distinction between essence and existence have? None, as Averroes was to show. This seems to be the basic weakness of Avicenna's position as regards *esse*. There is no doubt that he considered *esse*, not as a predicamental, but as a predicable accident.[8] But he never worked out the transition from the logical to the ontological order. Nonetheless, his notional analysis of necessity and possibility is related to real existence; but it is at this point that Avicenna slips from the *modus praedicandi* to the *modus essendi*. In other words, he regarded the logical possibility of finite things as forming real composition with their essence.

In conclusion, there are, in Avicenna's theory of emanation, at least five points deserving attention: (1) the distinction of essence and *esse*; (2) the nothingness of essence, by itself, thus emphasizing the importance of the existential order; (3) the two orders of causality: the creative and the predisposing; (4) the eternal and necessary creative flux; (5) the limitation of God's immediate influx to the First Intelligence, and so the subjection of the world to astral necessity.

The independence of Avicenna's speculation, and its reliance on Greek

[6] *Metaphysica*, VIII, 5, fol. 99v.

[7] J. Paulus, *Henri de Gand, essai sur les tendences de sa métaphysique* (Paris: J. Vrin, 1938), pp. 274–77. Although Avicenna starts his metaphysics with an analysis of the notion of being, we must not overlook his frequent references to the allegory of *The Floating Man*, cf. *De Anima*, V, 7, fol. 27v, which shows the possibility of an experience of pure existence, at least of one's own existence. See, A. Goichon, *La philosophie d'Avicenne et son influence en Europe médiévale* (Paris: Adrien-Maisonneuve, 1951), pp. 21–23, and *La distinction de l'essence et de l'existence d'après Ibn Sina*, pp. 13–15; E. Gilson, "Les sources gréco-arabes de l'augustinisme avicennisant," in *Archives d'histoire doctrinale et littéraire du moyen age*, IV (1929-1930), pp. 40–42.

[8] T. O'Shaughnessy, "St. Thomas' Changing Estimate of Avicenna's Teaching on Existence as an Accident,' in *The Modern Schoolman*, XXXVI (1959), 253–57.

thought rather than on the Koran, brought about a not unexpected reaction. But it is the source of this reaction which is interesting. Algazel had given great promise as a philosopher. But, when only 33, and apparently at the height of his career, he was afflicted with a severe crisis of doubt regarding everything. He afterwards attributed his cure to God, and turned to a life of Sufism. Some ten years later, at the request of a disciple, he composed his principal work, the *Destruction of Philosophers.* He first presented the metaphysics of Alfarabi and of Avicenna in an introduction, the *Intentions of the Philosophers,* and then attacked them in the *Destruction.*[9] It was Avicenna's necessitarianism which bore the brunt of the attack on philosophy. In defense of the faith, Algazel reaffirmed divine freedom in creating, creation *ex nihilo* and in time, and divine providence extending to all things. His method was to confront the conflicting opinions of philosophers on those points of faith and to show their inability to reach certainty on basic issues. Algazel died in 1111, and, incidentally, his work marked the decline of Arab philosophy in Persia and the East.

2. *Averroes*

Some seventy years later, the "Glory of Cordova" rose to the defense of philosophy. Averroes had already been engaged in ridding Aristotelianism, which for him was philosophy, of Neo-Platonic influences, especially those stemming from the cosmology of the *Timaeus* and of the *Theology of Aristotle.* Averroes considered Avicenna's system as the repository of most of the violence done to Aristotelianism, and the source of the trouble was that Avicenna had considered *esse* as accidental to the quiddity of things.[10] Time and again, Averroes touched on this distinction of essence and existence in his *Epitome* and his *Metaphysics.* Most of the passages of the *Destruction of the Destruction of Philosophers* are transcriptions of the *Epitome* or of the *Commentary on the Metaphysics.*[11] Averroes goes beyond Algazel to attack Avicenna's theory of being which was the basis of Algazel's criticism of philosophers:

[9] For the history of the *Maqacid* and the *Tahafut,* cf. B. Zedler, *Averroes' Destructio Destructionum Philosophiae Algazelis* in the Latin Version of Calo Calonymos, ed. B. Zedler (Milwaukee: Marquette University Press, 1961), pp. 5–9. For Algazel's criticism of Avicenna's theory of being, cf. fol. 33vb–34ra, pp. 248–49; fol. 34vb–35ra, pp. 254–59.

[10] *Averroes' Destructio Destructionum,* fol. 34va, p. 252: Nam ipse [Algazel] existimat quod entitas, scilicet quod res sit, est quid additum quidditati extra animam, et quasi accidit ei. . . . Quod autem esse apud Avicennam sit accidens eveniens quidditati, id indicat sermo Algazelis hic.

[11] Cf. M. D. Roland-Gosselin, *Le "De Ente et Essentia,"* pp. 157–59.

And this fellow has based his criticisms on the opinion of Avicenna which is altogether false.[12]

Averroes never hesitated; he consistently rejected any real distinction between the essence of actual things and their existence. And he was sure that he was following Aristotle. Unless *esse* is identically the substance of things, things have no essence. All his life, Aristotle had fought the mistake of hypostatizing forms and giving them reality apart from individuals. The real distinction of *esse*

is no different from Plato's doctrine of Forms outside of matter. And that is what Avicenna proposes!

Averroes was keenly aware of the importance for philosophy of avoiding any mistake in one's starting point. To make existence an accident in things is to deny them reality. As Aristotle has shown,

"One man" and "man" are the same thing, and likewise "an existent man" and "man," and the doubling of terms in "one man and an existent man" adds nothing to the meaning; it is clear that the being of man and his unity are not separate either in coming-to-be or ceasing-to-be; similarly "one existent man" adds nothing to "existent man," and so it is obvious that the additions in these cases do not change the meaning, and unity is nothing apart from being; and if, moreover, the substance of each thing is one, essentially and not accidentally, it is by its very nature something that is.[13]

If we grant that there are only two realms of being, the real and the intentional; if we further grant that only individuals exist in reality and that abstractions or universals, as such, do not subsist in themselves but exist only in knowledge; then – to be is to be an individual, a τί ἐστι καὶ τόδετὶ. As Averroes comments, to say "This substance is," is to say "This is a substance"; and to maintain that the being of a substance is accidental to it is to maintain that a substance is not a subsisting being. What could *esse* add to a real thing? How can one distinguish between the being of a thing and what that thing is?

One could not miss the Question of Being more completely. We have already met with a similar view in the case of the materialist for whom to be is to be a particle or complex of particles. The answer is that *esse* does not add anything to real individual things or to particles, but that *esse* is the exist-

[12] *Averroes' Destructio Destructionum*, fol. 34va, p. 252.

[13] *Metaphysics*, IV, 2, 1003 b 26–33; cf. J. Tricot, *Aristote, La Métaphysique* (Paris: J. Vrin, 1948), I, pp. 112–13; W. D. Ross, *Aristotle's Metaphysics* (Oxford: Clarendon Press, 1958), I, pp. 257–58.

ential act or actuating principle that makes individual things and particles real. For Averroes, *esse* means exclusively essence, as in the expressions τὸ ἀνθρώπῳ εἶναι or τὸ ἀγαθῷ εἶναι, the *hominem-esse* or *bonum-esse* of the scholastics. From this essentialist viewpoint. Avicenna's distinction is nonsense. To Averroes' way of thinking, it was returning to the atomicity of the Mutakallimum. For him, as for Aristotle, a thing is posited in being by the act which constitutes it what it is, i.e. by its form.

Averroes believed that he could detect a double confusion on the part of those who made existence an accident in things. The first source of error is a linguistic confusion. The Arab word for an existent is the past participle [mawjūd] of the verb *to be*. Popularly, the word is often used to mean that something has been found by accident. By extension, it also means something misplaced or there-by-accident. Now, some have been duped by this popular usage into regarding a thing's being as accidental and have concluded that existence is an accident in things.. Such is the mistake made by Avicenna when he says that *esse* is an accident superadded to the essence of a thing: *quod esse sit accidens eveniens quidditati.*

> This is a notorious instance, in philosophy, of uses by metonymy of the term *being*. For, the ordinary man's understanding of the term is very different from the meaning given to it by those who regard existence as an accident. In everyday use, we say that something is in something else to indicate that it is lost or somehow concealed. From this, some have mistakenly concluded that being is an accident in things, and not their essence, because they have not paid attention to the shift in meaning of the term *being* as used in ordinary conversation. But, as philosophers, we cannot accept that.[14]

But this linguistic clarification of Averroes is beside the point. The real issue involves two metaphysics of creation based on two different views of being-as-such. Avicenna's theory of being rests on the accidentality of *esse* with respect to essence considered in itself – the so-called indifference of nature-as-such as regards existing or not existing – and consequently the need of creation to account for the existence of things. For him, creation is the emanation of *esse,* through cognitional overflow, actualizing the infinity

[14] Averroes, *Aristotelis metaphysicorum libri XIII cum Averrois Cordubensis in eosdem commentariis et Epitome* (Venetiis: apud Juntas, 1562), Vol. VIII, *Epitome,* I, fol. 357vm–358ra. Istae ergo sunt famosiores res de quibus dicitur nomen entis in philosophia; et est de nominibus transumptivis, quia id quod a vulgo significatur per ipsum est aliud quam id quod significatur in hoc loco; nam plebs indicat aliquid inesse alicui rei, cum dicit invenitur perditum, tandem apud eos significatur aliquid inesse subjecto, licet non declaraverint ipsum. Hinc est quod aliqui existimaverunt ipsum significare accidens aliquod in re existens, et non ejus essentiam, quia illud apud plebem est de nominibus denominativis. Sed nos non debemus hoc recipere.

of possibles throughout eternity. In any actualized essence, *esse* is the inseparable concomitant of essence emanating from the Necessary Being. But, for Averroes, creation is the priming of the causal processes of the universe which educe forms from the eternal and unproduced potential, prime matter. As we shall presently see, Averroes understood by creation the dependence of the everlasting cycles of generation and corruption on the attraction exercised by the Prime Mover. And the reason is that, for Averroes as for Aristotle, being-as-such is being *qua* substance. The issue, therefore, is between the primacy of *esse* or the primacy of form.[15]

The second source of error, according to Averroes, comes from a confusion of logic and metaphysics. It is a case of mistaken identity:

> For the term *being* has two meanings: it may signify the truth of a proposition [as opposed to error] or it may signify real being as opposed to non-being. In this second sense, being is a quasi genus dividing into the ten categories, so that we say that substance is a subsisting being and that the accidents are determinations in subsisting being. But all ten categories equally share in being as the truth of a proposition.[16]

If *esse* were a real accident, Averroes continues, it would belong to one of the nine categories of accidents. Where does it fit? It is evident that *esse* cannot be an accident in things. On the other hand, the statement "Some substance is," may be either true or false. The verb *is* does not express a real accident residing in a subject, but a factual truth if there is a substance or a mistake if there is no substance. It is the truth of the statement which is accidental or, as logicians say, in contingent matter. To make existence a real accident in things is to mistake the mode of predication for a mode of being, that is, to confuse logic and metaphysics.[17]

It must be granted that Averroes, on this second point, scores against Avicenna. Insofar as Avicenna did not develop a metaphysics of existential act, but left *esse* as a predicable accident, he lay himself open to this criticism

[15] Cf. E. Gilson, *Being and Some Philosophers* (Toronto: Pontifical Institute of Mediaeval Studies, 1949), pp. 75–82.

[16] *Averroes' Destructio Destructionum Philosophiae Algazelis*, ed. B. Zedler, Disp. V, fol. 34va, p. 253: Nomen entis dicitur de duobus. Unum quidem de vero, et alterum de eo quod opponitur privationi. Et hoc est quod dividitur in decem praedicamenta, et est tanquam genus eis.... Et hoc modo dicitur de substantia quod sit ens ex se et de accidenti quod sit ens, ex eo quod reperitur in ente ex se. Ens vero, quod est de vero, communicant in eo omnia praedicamenta aequaliter. Cf. again *In Meta.*, V, 7, fol. 117rb; *Epitome*, I, fol. 358rb.

[17] *Averroes' Destructio Destructionum Philosophiae Algazelis*, ibid., p. 253: Ens quidem quod est verum, est quid in intellectibus, et est esse rei extra animam, prout in anima.... Cum autem aliquis dicit quod esse est additum quidditati, et non constituitur eo ens per se, est sermo fallax nimis....

of Averroes. That the latter misrepresented Avicenna's view of the accidentality of *esse* does not affect the basic issue: there is no real distinction between *esse* and essence, unless *esse* is an existential act. The denial of the real distinction led Averroes to differ with Avicenna on the metaphysical structure of beings. Avicenna had considered all creatures as, by themselves, mere possibles which are necessarily posited in existence by a unique Necessary Being. His theory made necessity depend exclusively on an extrinsic cause. To Averroes, this idea was another instance of Platonic extrinsicism.[18] Following Aristotle, Averroes considered necessity as intrinsic to the nature of pure forms and not as a relation of causal dependence. Whatever is essentially act, without admixture of potency, is necessary and immutable; whatever is essentially composed of potency and act is contingent and corruptible. Necessity and contingency define kinds of beings. This change of perspective, from the existential to the essential, enabled Averroes to sweep away the theory of emanations and to return to a strictly Aristotelian view of reality. In this hylomorphic dualism, pure acts as well as prime matter are uncaused and eternal. The number of pure acts is calculated according to the number of spheres in astronomy; as for prime matter, since it cannot exist without form, its existence is assured by the eternity of the processes of generation and corruption.[19]

Averroes criticized, on Aristotelian grounds, the theologians of Islam, Judaism, and Christianity, the *Loquentes trium legum*, for holding creation *ex nihilo* at the beginning of time. There cannot be any coming-to-be without

[18] *Averroes' Destructio Destructionum*, Disp. III, fol. 25 ra, pp. 187–88: Quod autem est necessarium ex causa, non est necessarium ex se. Et nemo dubitat quod istae differentiae non sunt differentiae substantiales, idest dividentes substantiam, nec additae substantiae; sed sunt res negativae et relativae; sicut dicimus de re inventa quod non significat rem additam substantiae extra animam, eo modo quo dicimus de re quod sit alba. Et ex hoc erravit Avicenna.... For the source of the doctrine in Plato, cf. *Timaeus*, 41 a 7–9: Gods, children of gods, who are my works and of whom I am the maker and father, my creations are indissoluble, if so I will. All that is compounded can be resolved ... Wherefore, since you are creatures, you are not altogether imperishable, but you shall certainly not be dissolved.... Cf. Calcidius, *Timaeus*, II, CXXXIX (ed. J. H. Waszink, Corpus Platonicorum Medii Aevi, Vol. IV; Leiden: E. J. Brill, 1962), pp. 179–80. St. Thomas frequently refers to this view, e.g. In II *Sent.*, I, 1, 3, obj. 1; I, 50, 5, ad 2; *De Spir. Creat.*, I, arg. 18; *De Coelo et Mundo*, I, 12, lect. 29, n. 8.

[19] Cf. L. Kendzierski, "The Doctrine of Eternal Matter and Form," in *The Modern Schoolman*, XXXI (1953-1954), 171–83; M. Fakhry, *Islamic Occasionalism and Its Critique by Averroes and Aquinas* (London: Allen and Unwin, 1958), pp. 118–24; G. Jalbert, *Necessité et contingence chez saint Thomas et chez ses prédécesseurs* (Ottawa: Editions de l'Université d'Ottawa, 1961), pp. 37–42. The case of the celestial bodies seemed to make exception; but Averroes explained their immutability, as Aristotle had, by recourse to the doctrine of quintessence.

substrate, he argued, since the generation of anything is always the corruption of something else. An absolute coming-to-be is inconceivable: out of nothing, nothing comes. Time also is eternal, for the present "Now" is necessarily the link between the past and the future. Averroes was merely repeating Aristotle.[20] He advanced two reasons for the errors of the theologians: (1) to the uneducated, many things seem to come from nothing since their causes are hidden; and so some have mistakenly concluded that it must be possible for something to come from nothing. (2) They then confirm their original mistake by adding that the need for a subject matter to work on is the sign of imperfection on the part of the agent. However, Averroes remarks, to say that the impossible cannot be is no restriction on anyone's power.

These criticisms of Averroes were directed especially against the occasionalism of the Asharites. As a Moslem, Averroes accepted creation, but he gave it an Aristotelian interpretation. Explaining how the production of beings must be finally attributed to God, he wrote:

> According to the Philosopher, things whose being consists in the juncture of co-principles, such as matter and form, depend for their being on the author of this juncture, so that the cause of the union of the co-principles is cause of the being.[21]

Faithful to Aristotle's essentialism, Averroes could still regard creation as the ontological dependence of the universe on God as *Inventor of substances*. As Prime Mover, God is the universal cause of all that comes to be, and "on such a principle, the heavens and nature depend." [22] Averroes was confident that this explanation accounted for being-as-such.

The upshot of this discussion is that we are faced with two different interpretations of creation based on two different views of being. Avicenna, in order to preserve contingency in the universe, was led to distinguish existence from essence in all beings but God. This distinction made him see the need of creative causes, over and above moving causes, precisely to account for existence. But his explanation of creation, geared to the *Theology of Aristotle*, could not do justice to his insight. His theory of emanations, as J. Paulus has pointed out, remained on the plane of noetic determinisms. Consequently, his God is a *Necesse Esse* whose aseity is assured by the fact

[20] *Phys.*, VIII, 1, 251 b 10–28; *De Coelo*, I, 12, 281 b 26 – 282 b 12; *De Generatione et Corruptione*, II, 10, 337 a 15 – 337 b 35.

[21] *Averroes' Destructio Destructionem*, Disp. III, fol. 24ra, p. 181; cf. M. Fakhry, *Islamic Occasionalism*, pp. 129, 131, 138.

[22] *Meta.*, XII, 7, 1072 b 14.

that he has no essence, since essence would make him a possible-of-itself, and not by the fact that he is *Ipsum Esse* intensively considered.

Averroes blamed Avicenna for "listening to the theologians of our religion." He himself intended to return to pure philosophy which, for him, was historical Aristotelianism as he read it. Considering being as οὐσία, he reached a creator who is a Prime Mover. In his *Commentary on the Physics*, St. Thomas criticized Averroes' arguments against the Christian notion of creation and reproached him for staying on the level of particular coming-to-be and not considering the production of the whole entity of things *ex nihilo* which alone accounts for being *qua* being:

> Arguing from Aristotle's viewpoint, Averroes attacked the position which we hold on faith concerning creation. But he failed to see that, in treating of creation, we are not talking of physical processes precisely because he himself never saw beyond particular changes. Surely, if some particular being comes into existence by way of natural processes, it cannot come to be out of absolute non-being; but if the whole entity of a thing comes into existence, which is the coming-to-be of a being insofar as it is being, it must come out of absolute non-being, if we may speak properly of a coming-to-be in this instance, since the term is used equivocally.[23]

Is St. Thomas, in this passage, merely extending and prolonging Aristotle's explanation of efficiency beyond prime-moving to creation? It must be noted that, throughout this section of the Commentary, St. Thomas is only pointing out the ineffectiveness of the arguments of Averroes against the Christian notion of creation. St. Thomas is assuming revelation and arguing that, *secundum nostrae fidei sententiam*, creation is not a physical process. It is quite another matter to establish, independently and on metaphysical grounds, that creation, in the sense of the production of the whole entity of things, is necessary to explain being *qua* being. The theologian,

[23] *In Phys.*, VIII, lect. 2, nn. 1996–2006: Ex hac autem Aristotelis probatione, Averroes occasionem sumpsit loquendi contra id quod secundum fidem de creatione credimus.... Sed si quis recte consideret, ex simili causa ipse deceptus fuit ex qua causa nos deceptos arbritatur, scilicet ex consideratione particularium entium.... Si fiat aliquod particulare ens non fit ex omnio non ente; sed si fiat totum ens, quod est fieri ens inquantum est ens, oportet quod fiat ex penitus non ente; si tamen et hoc debeat dici fieri, aequivoce enim dicitur.... R. Jolivet, *Essai sur les rapports entre la pensée grecque et la pensée chrétienne* (Paris: J. Vrin, 1955), pp. 48–49, comments: "St. Thomas intends to prove that fidelity to Aristotelian principles leads inevitably to the distinction, which he here advocates, between generation and the production of the whole entity of beings. The reasoning of St. Thomas is simply an extension of Aristotelian doctrine. One might say that St. Thomas forges to the end of the trail opened by Aristotle." – Our contention is that the trail to creation is not along the path of essence, but along that of existence. It leads from existence as a fact to existence as existential act.

accepting creation on faith, can show that a certain account of the existence of beings is insufficient in the light of faith. He can argue that unless creation is understood as the total production of an entire being, the first cause will not be agent in an absolute sense, but cause only on transformations in a potential or of the ordering of a primitive chaos which it has not made. The theologian can equate the "coming-to-be of the whole being" with the "coming-to-be of a being *qua* being." But the fact of creation, the theologian holds on faith. The metaphysician has not, as metaphysician, this light of the theologian. In metaphysics, it is not enough to remark that creation deals with the production of being *qua* being, for this is the whole issue. Averroes, following Aristotle, believed that he had explained being *qua* being, according to his understanding of the term *being*. Can one go beyond Aristotle's solution while remaining within his formal perspective? Is it possible to prove creation, in the sense of the production of the whole entity of a being, in a metaphysics of form or essence? Is there any intrinsic reason why there need be any production of the whole entity of anything or *creation ex nihilo,* except to explain existence? This is what the term *creation out of nothing* tries to convey: in order to account for beings, absolutely speaking and without supposing anything, one must pass from the order of change and mutations to that of existence. For change requires a potential; but creation, giving *esse,* thereby produces that which receives *esse,* and so does not presuppose any substrate.[24] Creation means passing from the formal order to the existential order; from *esse actu per formam,* which is Aristotelian, to *actus essendi,* which is Thomistic; from being *qua* subsistent to being *qua* existing. As we shall see in the next chapter, no form can create or serve as instrument in creation; the creator, in the full sense of the word, must be *Ipsum Esse* and his proper effect must be *actus essendi*. For the time being, we may quote this conclusion of S. Mansion:

> A Prime Mover, as conceived by Aristotle, cannot fulfill the role of creator. Something more is needed. And the reason for the insufficiency of the Prime Mover can perhaps be traced to the unredeemed Platonism of Aristotle in spite of his severe criticism of the theory of Forms. Aristotle's notion of *ousia* is too much like a Platonic Form to represent adequately the Transcendent Being.

[24] The case of Siger de Brabant is instructive. In *Quaestiones in metaphysicam* (ed. C. Graiff, Louvain: Institut Supérieur de Philosophie, 1948), qu. 7, pp. 16–22, Siger confessed that he could not make any sense of a real distinction between *esse* and essence. Later, qu. 18, p. 153, Siger was exercised by the objection that an absolute beginning or creation *ex nihilo* is a contradiction in terms. Siger claimed that Aristotle nowhere proves that it is a contradiction, but merely asserts that it is. Siger believed that Aristotle was wrong from the viewpoint of Christian faith; but, speaking as a philosopher, he admitted that he could not meet the objection.

So long as being, in its primary instance, is conceived as that which is identically *what it is,* so long as essence or form is viewed absolutely as perfection, and not as *limit* as well, the range of metaphysics is restricted. Even though enriched and fecundated by the notion of act, the highest form would still be unable to account for reality in its totality and diversity. In the hands of Aristotle, the metaphysics of Form yielded perhaps as much as it could. The metaphysics of act, however, never gave its full measure. It is this aspect of Aristotelianism which St. Thomas exploited in his metaphysics of *actus essendi,* and which led him to the notion of God in whom essence and existence coincide in simple actuality.[25]

If we may speak of the prolonging of Aristotelian principles by St. Thomas, it should be in line with the notion of act. This is what a genetic method does. First, we start with Aristotle's definition of motion or change as an existential act. If change is an imperfect existential act, then *esse,* the result of change, is also an existential act:

> as motion is the act of a body in motion, so is existence the act of an existent as such.[26]

Next, we notice that Aristotle extends the notion act or ἐνέργεια to immanent activity. This leads him to regard the First Mover as Self-subsisting Consciousness.[27] Here again *act* breaks from the order of forms into life as pure immanence or pure actuality of thought. Although understood in an intellectualistic manner, this divine actuality transcends all potency. Act is self-subsisting ἐνέργεια. This is the closest that Aristotle comes to considering act intensively. His view transcends the order of forms – and so the order of change – to reach pure actuality in the order of immanent activity.

Combining these two aspects of act – the existential character of act revealed in motion and the intensive character of act revealed in immanence – it becomes imperative to conceive existence in terms of *actus essendi.* For, in the line of act,

> it is necessary to seek, beyond physical changes which educe forms from the potentiality of matter, an absolute origin of things whereby *esse* is bestowed on all things in the universe by the First Being who is his *esse.*[28]

[25] S. Mansion, "Les positions maîtresses d'Aristote," *Aristote et saint Thomas d'Aquin* (Louvain: Publications Universitaires, 1957), pp. 66–67.

[26] In I *Sent.*, XIX, 2, 2 (ed. Mandonnet, I, p. 470): sicut motus est actus ipsius mobilis, inquantum mobile est; ita esse est actus existentis, inquantum ens est.

[27] *Meta.*, L, 9, 1074 b 34.

[28] *De Subs. Sep.*, 7, n. 49: Oportet igitur supra modum fiendi quo aliquid fit, forma materiae adveniente, praeintelligere aliam rerum originem, secundum quod esse attribuitur toti universitati rerum a primo ente quod est suum esse.

To conclude: if the two aspects of total causality, (1) the production of the whole entity of things from nothing or (2) the bestowing of *esse*, are the same, materially considered, nevertheless the formal aspect of creation is the participation of existential act. The production of the whole entity is an essentialist viewpoint; the bestowing of *esse* is existentialist. St. Thomas indifferently used one or the other approach, depending on the context or the sources that he was explaining. Thus, in order to show the total dependence of finite beings on a creative cause, he will argue that the creator is the cause of *esse* and so of all the determinations which contract and diversify *esse*, including matter. On other occasions, in order to show the insufficiency of physical agency to account for being-as-such, since physical agency produces only transformations, he will stress the need of a cause of the whole entity of things. To the theologian, either way is permissible. But, in metaphysics, only the existential account will do. The medieval controversy over the possibility of an eternal universe, which we are about to consider, confirms this conclusion.

CHAPTER IV

CREATION AND EXISTENTIAL ACT

The impact of Greco-Arabian philosophy on the West dominates the intellectual climate of the thirteenth century.[1] For some time it had been rumored throughout Europe that, long before the advent of Christianity, Aristotle had expounded a complete system of sciences in the pure light of natural reason. Here was no faith seeking understanding, as in St. Augustine and St. Anselm, but an autonomous scientific account of the universe that rivalled anything written by the saints. Its logical completeness made it look like the last word in human wisdom. And it offered what the West completely lacked: a vast body of natural sciences. In the newly unified Christendom, this idea of explaining reality without reference to revelation sounded as novel and alarming as the idea of ruling society without reference to divine law. Freedom of investigation as well as liberty of princes were on the same plane: they were liberties allowed within the Church, liberty to exercise reason and liberty to take administrative measures, so long as they did not conflict with Christian principles. From the earliest days of the Church, theology had made abundant use of Platonism; the medieval discovery of the physical treatises of Aristotle raised a different kind of problem. Here was the first contact of the Christian mind with what passed for science. Could peripatetic philosophy be harmonized with the faith? To complicate matters, Aristotle's text came in partial translations together with Arab commentaries and Neo-Platonic works, such as the *Liber de Causis,* which circulated under Aristotle's name, the *De Unitate,* wrongly attributed to Boethius, and the *De Fluxu Entis,* mistakenly believed to be Avicenna's.

[1] F. Van Steenberghen, *Aristotle in the West,* tr. L. Johnston (Louvain: Institut Supérieur de Philosophie, 1955), pp. 59–114; M. Grabmann, *I divieti ecclesiastici di Aristotele sotto Innocenzo III e Gregorio IX* (Rome: Saler, 1941), pp. 5–69; E. Bréhier, *Histoire de la philosophie,* T. I, pp. 634–88; B. Gillon, "La lutte pour l'aristotelisme modéré," in *Dictionnaire de théologie catholique,* XV, 654–71; P. Glorieux, "La littérature des Correctoires," in *Revue thomiste,* XI (1928), 69–96.

Something of the dismay of the theologians, especially of the Augustinian tradition, is reflected in this passage of William of Auvergne:

> As for the Christian people, wholly dedicated to virtue and holiness and the service of the Creator, it has been occupied very little with philosophy, save when the perversity of heretics and the objections of fools have compelled it to defend its religion and its faith and to destroy the doctrines that are opposed to its salvation and contrary to the honor of God. The men of this religion have been concerned with their souls and have not troubled themselves about the souls of the heavens. It has seemed to them that, from the point of view of their religion and their eternal salvation, there was neither profit to be gained from knowledge about these souls nor loss from ignoring them. Whether the world may be a single animate body or not, whether the entire heaven should be an animate being or the different heavens animate beings – these are questions that the Christian people have regarded with horror and treated as monstrous. It is plunged in astonishment by this discussion which has hitherto been entirely unknown to it, and in which it sees a novelty that does not concern it at all.[2]

Besides world-souls, the most discussed questions were those of the separate Agent Intellect and of creation in time. Both works of St. Thomas on these points, *De Unitate Intellectus contra Averroistas* and *De Aeternitate Mundi contra murmurantes*, witness to the moods of the times. As regards creation, which alone concerns us here, the theologians charged the philosophers with (1) failing to reach any genuine notion of creation, (2) maintaining the eternity of the universe, (3) advancing theories of necessary emanations of being through intermediaries.[3] St. Thomas did not quite share the views of his contemporaries on these points, but, more important, he disagreed with them on philosophical grounds. It is these grounds that we

[2] William of Auvergne, *De Universo*, P. I, III, c. 31 (*Opera Omnia*, Aureliae: F. Hotot, 1674, T. I, p. 805).

[3] Cf. A. Forest, *La structure métaphysique du concret selon saint Thomas d'Aquin* (Paris: J. Vrin, 1956), pp. 46–71; D. Losa, "S. Thomas et le problème de la création," in *Revue thomiste*, L (1950), 157–71. Alexander of Hales, *Summa Theologica* T. II, tr. 1, sect. 1, qu. 2, c. 6 (ed. Quaracchi, p. 37), quotes Aristotle and Avicenna on creation; but his reference to Aristotle is the corrupt text, as in the translation of Gerard of Cremona, of the *De Coelo*, I, 1, 268 a 14: Per hunc numerum [ternarium] adhibuimus nos ipsos magnificare Deum Creatorem eminentem proprietatibus eorum quae sunt creata. Alexander concluded that some philosophers had had a notion of creation. St. Albert disagreed, cf. *In II Sent.*, D. I, A, a. 8 (ed. Borgnet, XXVII, p. 22), "no philosopher had any idea of creation, unless one or another chanced to read some passage of our sacred writers; but no philosopher ever instituted a demonstration of creation." St. Bonaventure, In II *Sent.*, D. I, P. I, a. 1, qu. 1 (ed. Quaracchi, p. 16), regarded creation as a matter of revelation: "This truth, so clear and certain to the least of the faithful, remained hidden from the learning of the philosophers." William of Auvergne attacked especially the Avicennian theory of necessary emanations; but he praised Avicebron for his views on free creation through the Word, and believed that he must have been a Christian, cf. *De Universo*, P. I, I, 26 (Opera Omnia, p. 621).

must now investigate in order to discover what link can be established between creation and existential act.

1. The Long Trek

On at least six occasions, in various works, St. Thomas described the slow progress of philosophers in their search for the ultimate source of reality.[4] St. Thomas never hesitated. From the time of the *Commentary on the Sentences*, he resolutely stated that creation can be demonstrated philosophically and that some philosophers did offer proofs of "a universal cause of the being of things." [5] The progress is usually described as moving from accounts of accidental changes to explanations of substantial transformations and finally to the production of the whole entity of things:

The ancient philosophers only gradually and, as it were, step by step advanced in the knowledge of truth. At first, being rather inexpert, they believed that all beings were sensible bodies. And so, those among them who admitted movement did not consider it except according to certain accidents, for instance, according to rarefaction and condensation, through union and separation. And supposing, as they did, that corporeal substance itself was *uncreated*, they assigned certain causes for these accidental changes such as friendship, discord, mind, or something of that kind.

An advance was made when philosophers understood that there is a distinction between substantial form and matter, which latter they however held to be *uncreated*, and proceeded to explain transmutations in bodies in terms of essential forms. These transmutations they attributed to certain *more universal causes*, such as the ecliptic according to Aristotle or the Ideas according to Plato. However, we must take into consideration that matter is contracted by its form to a determinate species, just as a certain substance belonging to a certain species is contracted by a supervening accident to a determinate mode of being as, for instance, man by whiteness. Each of these opinions, therefore, considered being under some particular aspect, namely, either as *this* being or as *such* a being; and so they assigned particular efficient causes to things.

[4] *Contra Gentiles*, II, 16 and 37; *De Potentia*, 3, 5; I, 44, 2; *In Phys.*, VIII, 2, n. 2007; *De Substantiis Separatis*, 7, 49–53; *Compendium theologiae*, c. LXVIII. For comparative studies of the texts, cf. J. de Finance, *Etre et Agir*, pp. 126–27; F. Giardini, "Similitudine e principio di assimilazione," in *Angelicum*, XXXV (1958), 309–10; A. Pegis, "A Note on St. Thomas, Summa Theologica, I, 44, 1–2," in *Mediaeval Studies*, VIII (1946), 159–68; A. Hayen, *La communication de l'être*, pp. 38–43.

[5] In II *Sent.*, I, 1, 2, c (ed. Mandonnet, II, pp. 17–18): Respondeo quod creationem esse (1) non tantum fides tenet, sed etiam ratio demonstrat. . . . (2) Sic philosophi creationem posuerunt. . . . (3) Si autem accipiamus oportere ad rationem creationis, ut scilicet etiam duratione res creata prius non esse quam esse habeat, ut dicatur esse ex nihilo, quia est tempore post nihil, sic creatio demonstrari non potest, nec a philosophis conceditur, sed per fidem supponitur.

Finally, a few advanced further and rose to the consideration of being-as-being and assigned a cause to things, not only according as they are these and such, but according as they are beings. Therefore, whatever is a cause of things considered as beings must be the cause of things, not only according as they are so by accidental forms, nor only according as they are *such* by substantial forms, but according to all that belongs to them in any way whatever. And thus it is necessary to say that even prime matter is created by the universal cause of beings.[6]

The various texts present a well-known problem concerning the place assigned by St. Thomas to Plato and Aristotle in this progress of human reason towards the explanation of being-as-being. The *De Potentia,* the *Commentary on the Physics,* and the treatise *On Separate Substances* place them in the final stage and credit them with giving an explanation of being-as-being; whereas the *Summa Theologiae* puts them in the second stage and indicates their acceptance of eternal, uncaused matter as the reason for their failure to reach the idea of creation. It is inevitable, in a metaphysics of Form, that matter will prove embarrassing and, as Sartre would say, *de trop.* It is there, but cannot be accounted for. Matter being matter, since it does in fact exist, it will endure everlastingly throughout all the processes of nature. But need it be? It is existence which has to be justified. Plato and Aristotle had to regard matter as eternal and unproduced because, in their purview which equated intelligibility with form, existence is taken for granted but never explained. And so they failed to rise above predicamental causality to transcendental causality.

And yet, as E. Gilson remarks, "*esse* has both a limited meaning and a broad one. Its strict and properly Thomistic meaning is 'to exist.' Its broader, Aristotelian meaning is substantial being . . . St. Thomas could say, without contradicting himself, that Aristotle came to acknowledge a first cause of *totius esse,* in the sense of substantial being, and also that he never came to accept the notion of God as creator, that is, cause of existential being." [7]

[6] This is the account of the *Summa Theologiae,* I, 44, 2. St. Thomas gives no clue to the identity of the philosophers who reached the idea of creation. E. Gilson, *The Christian Philosophy of St. Thomas Aquinas* (New York: Random House, 1956), p. 132, believes that St. Thomas may have had in mind certain Platonists praised by St. Augustine in *De Civitate Dei,* VIII, 4 (P.L. 41, 228). C. Fabro, *Participation et causalité,* p. 368: "Who are the *aliqui*? St. Thomas does not tell us in this article; but we know for sure, from his other writings, that they are the Platonici or Neo-Platonists." J. de Finance, *Etre et Agir,* p. 127, thinks that St. Thomas was referring to Proclus, to the author of the *Book on Causes,* and to Avicenna whom he always regarded as a philosopher.

[7] *The Christian Philosophy of St. Thomas Aquinas,* p. 461.

That would seem to be the source of the ambivalent attitude of St. Thomas towards Aristotle. Any production of a new being brings something into existence, and, certainly, the total production of a whole being from nothing will give existence, whether considered as the mere fact of having been produced or as *actus essendi*. However, St. Thomas was familiar with a long theological tradition which defined creation, not by explicitly emphasizing its existential character, but rather by stressing the totality aspect of the causality of the creative act. Creation is then defined as:

> the emanation of the whole being of a thing,
> the production of something from nothing,
> the giving of its whole substance to a thing,
> the production of a whole subsisting being,

which definitions are perfectly compatible with a metaphysics which denies that *esse* and essence are really distinct.[8] What is more, such definitions agree with the biblical idea of creation as expressed, for instance, in the second Book of Machabees.[9] St. Thomas all his life believed that such a view was contained, in germ at least, in Aristotle; for he always maintained that Aristotle's Prime Mover was efficient as well as final cause of the universe.[10] That St. Thomas was mistaken on this point is incidental, although it explains his interpretation of Aristotle. But the problem is deeper. The two perspectives – the essentialist and the existentialist – are involved in this question. Is it possible to prove a cause *totius esse*, in the sense of substantial being, in an essentialist perspective? St. Bonaventure stated the issue very clearly:

> Both the *sancti* and the philosophers agree that material things need a first principle because of their multiplicity, their mutability, their order, their limitations... For multiplicity depends on unity, change on the immutable, order

[8] Such definitions are subscribed to by men like Henry of Ghent, Duns Scotus, and Suarez who rejected the real distinction of *esse* from essence. Cf. J. Paulus, "Les disputes d'Henri de Gand et de Gilles de Rome sur la distinction de l'essence et de l'existence," in *Archives d'histoire doctrinale et littéraire du moyen âge*, XIII (1940-1942), 351–58; Duns Scotus, *Ordinatio*, I, Disp. 36, qu. unica (Opera Omnia, ed. C. Balic, Civitas Vaticana: Typis Polyglottis, 1963, pp. 276–83); F. Suarez, *Disputationes Metaphysicae*, Disp. XX, sect. 1, n. 29 (ed. Vives, XXV, p. 753), and Disp. XXI, sect. 1, n. 6, p. 787.

[9] 2 Machabees, 7:28, "Look upon heaven and earth and all that is in them, and consider that God made them out of nothing...."

[10] *In de Coelo et Mundo*, I, 8, n. 14: Est attendendum quod Aristoteles hic ponit Deum esse factorem coelestium corporum, et non solum causam per modum finis, ut quidam dixerunt. Cf. *In Meta.*, II, lect. 2, n. 295; VI, lect. 1, n. 1164; *De Substantiis Separatis*, c. III, n. 15; VIII, n. 59; *In de Causis*, prop. 18a (ed. Saffrey, p. 103); In II *Sent.*, I, 5, ad 1 in contra (ed. Mandonnet, II, p. 38).

on a first, limitation on a perfect being. Granted, then, that there is a first principle of things, the queston is: is this first principle total cause of the whole being of things, of their matter as well as of their form, or only of their form?[11]

In other words, does the universe witness to a Prime Mover or a First Cause or a Necessary Being only, or does it point to a Creator? St. Bonaventure honestly doubted; and, after passing in review the Pre-Socratics, Plato, and Aristotle, he concluded:

Where the learning of the philosophers failed, the sacred scriptures enlighten us by teaching us that all things, according to their whole being, have been created in existence.[12]

We have only to consider the first three ways of demonstrating God's existence in the *Summa Theologiae* to appreciate the problem. They all start from motion or change: there is motion in the universe as our senses testify; there are chains of causally dependent agents; there are beings that come into existence and pass away . . . Is the Prime Mover or the First Cause or the Necessary Being, reached in the conclusion, total cause of the whole entity of the effects? Not in Aristotle, nor in Averroes, nor, it may be added, in anyone else who sticks to their perspective. Formal perspectives are limiting.[13] If the premises start from motion and change, which presuppose a potential, the conclusion will never reach a total cause. To be total, a cause must give existence.[14] We do have in St. Thomas another set of definitions of creation which express this existential dimension of reality, assumed but never explicitly asserted, by the first set. Creation, accordingly, is

issuing *esse*
giving each and all their *esse*
the outflow of *esse* [*influentiam essendi*]

[11] In II *Sent.*, D. I, P. I, a. 1, qu. 1 (ed. Quaracchi, p. 14): Cum constet secundum sanctos et philosophos, quod omnes res mundanae habuerint principium productivum tum propter rerum varietatem, tum propter rerum mutabilitatem, tum propter rerum ordinem, tum etiam propter imperfectionem. . . . Multitudo enim ortum habet ab unitate, et motus sive mutabile ab immutabili, et ordo ad primum et imperfectum ad perfectum originaliter reducuntur; ideo hoc supposito, est quaestio: utrum res sint productae omnio, h.e. secundum principium materiale et formale, an tantum secundum alterum principiorum.

[12] Ibid., p. 17: Ubi autem deficit philosophorum peritia, subvenit nobis sacrosancta scriptura quae dicit omnia esse creata et secundum omne quod sunt in esse.

[13] St. Thomas, *In Phys.*, VIII, lect. 21, n. 2453: sic intelligenda est conclusio quemadmodum sequitur ex praemissis. . . .

[14] In I *Sent.*, 36, 1, 1, c: Deus non [est] tantum causa esse rerum sed omnium quae in rebus sunt. . . . *De Potentia*, 3, 1, ad 17: Deus simul dans esse, producit id quod esse recipit; et sic non oportet quod agat ex aliquo praeexistenti. Ibid., 5, ad 2: Hoc ipso quod quidditati esse attribuitur, non solum esse, sed ipsa quidditas creari dicitur.

causing an effect according to its *esse*
producing something in *esse* according to its whole substance
without presupposing anything
a certain accession, so to speak, to *esse*
the *per se* cause of *esse*.[15]

Admittedly, the two aspects of creation – the production of the whole entity of things and the cause of an effect according to its *esse* – blend in reality. Hence, the theologian, who knows from revelation the total causality of God, can equate *fieri totius entis* and *fieri ens inquantum ens*. He can move in what looks like a continous line from accidental changes to substantial changes to the production of the whole entity of things. But, in so doing, he is passing without warning from the order of formal actuations of an existing substrate to the absolute giving of existence. In metaphysics, only the existential approach calls for a total cause. To account for existents as such, one must pass from the order of transformations to the order of existence. It is by training the mind's eye on existence, and by keeping existence in focus, that one is led to seek an explanation of reality beyond process and actuations.

In the light of these considerations, it is already possible to formulate the two conclusions of this study:

(1) Unless *esse* is distinct from essence, and unless *esse* is what has to be accounted for, there is no more reason for a universal cause of things under the aspect of being – *sub ratione entis seu inquantum entia* – than for universal causes of things under the aspect of corporeity, animation, sentiency, rationality, or any degree of ontological perfection which the mind distinguishes in things. Such reasoning, e.g. in the *Book on Causes*, is simply a Neo-Platonic version of the *via Platonica*. It regards whatever is common to various effects as due to universal causes. Consequently, it posits both the multiplicity and the separate subsistence of universal causes.[16] The being of things is indistinct from its essential degrees of perfection; unless, by being,

[15] Cf. In II *Sent.*, I, 1, 2, c, and ad 1; *De Potentia*, 3, 1, c; 3, 4, c; 3, 7, c; I, 45, 2, ad 2; *De Substantiis Separatis*, 7, n. 50: Oportet igitur, supra modum fiendi, quo aliquid fit per mutationem vel motum, esse aliquem modum fiendi seu originem rerum absque omni mutatione vel motu per influentiam essendi.

[16] St. Thomas rejected both the grounds and the conclusion of such a world-view as involving (1) the transposition of abstractions into reality; (2) the multiplicity of forms in one and the same natural unit; (3) a corresponding plurality of separate Form-causes often regarded as deities or *dii providentes*, cf. R. Henle, *St. Thomas and Platonism*, pp. 355–58, 413–17. To that ontological pluralism, St. Thomas opposed the simplicity of *Ipsum Esse*, including within himself all absolute perfections, and the composition in others of participated *esse* and limiting essence.

we mean *esse* or *actus essendi*. Credit is due to Avicenna for transposing the problem from the essential into the existential order, and to St. Thomas for completing the transposition by making *esse* an existential act.

(2) For, unless *esse* is an existential act, there is no metaphysical ground for a real distinction between *esse* and essence. Considered purely from a logical viewpoint – as a predicable accident of finite beings – *esse* can stand for anything from "to be the value of a variable," at one end of the spectrum, to the position of a thing outside its causes, the actuality of an essence, *das Nichts*, or existential act, at the other end of the spectrum.[17] What has to be kept in perspective is the ontological primacy of *esse*. Then *esse* appears as the basic reality which all things share in various degrees. Essence is the intrinsic measure, in a being, of its *esse*. Esse is act; essence, potency. It is this existential viewpoint that the medieval controversy about the eternity of the universe brought to the forefront.

2. *The Controversy over the Eternity of the World*

Viewed exclusively from the angle of matter and change, the eternity of the universe is a foregone conclusion. For matter cannot come-to-be insofar as it is matter. As such it is the enduring substrate of all substantial changes. Every coming-to-be involves corruption of a previous composite, so that the processes of generation and corruption are everlasting. In a passage of his *Commentary on the Metaphysics*, destined to become controversial in the thirteenth century, Averroes had reaffirmed this thesis against the Mutakallimum doctrine of creation in time.[18]

To the majority of Christian theologians of the thirteenth century this acceptance of the eternity of the world was proof that the philosophers had never reached a true notion of creation.[19] The problem, as the theologians

[17] C. Fabro has waged a relentless campaign against the metaphysically bland expressions such as *esse actu, esse actualis essentiae, esse existentiae,* to characterize the *actus essendi* of St. Thomas. Cf. *Participation et caualité*, pp. 26–83, 280–315.

[18] Averroes, *In Meta.*, XII, 143ra. In a corresponding passage of the *Destructio Destructionum*, Disp. I, 12va (ed. Zedler, pp. 99–100), Averroes observes that the philosophers hold that the world is eternal, not *per se*, but *per accidens*, in the sense of continuous process. Hence the eternity of the world does not negate its being caused.

[19] C. de Calmtout, "Possibilité ou impossibilité du monde éternel," in *Etudes Franciscaines*, VIII (1902), 249–68, 348–74, 511-28; E. Longpré, "Thomas d'York et Matthieu d'Aquasparta: Textes inédits sur le problême de la création," in *Archives d'histoire doctrinale et littéraire du moyen âge*, I (1926-1927), 269–72; M. Gierens, *De Aeternitate Mundi Controversia* (Rome: Aedes Pont. Universitatis Gregorianae, 1933); C. Vollert, L. Kendzierski, P. Byrne, *St. Thomas Aquinas, Siger of Brabant,*

saw it, was: can caused beings have existence without beginning to be? The classical argument had been formulated by Alexander of Hales as follows:

> To the question: why cannot the world have existed from all eternity, since it depends only on the will of the Creator who could have willed it so, I answer that the notion of creature implies a being produced out of nothing, and that every caused being has existence after non-being. Hence, an eternal creature is a contradiction in terms.[20]

St. Thomas regarded this argument as a throwback to the verbal reasoning of the Mutakallimum which had been exposed and refuted by the philosophers.[21] Forced to clarify the philosophical notion of creation, St. Thomas used the same two points mentioned by the theologians, *fieri ex nihilo* and *esse post non-esse*, but he transposed them from the order of motion and time and made them components of the metaphysical structure of things. In fact, so long as creation is viewed as a passage or transition from nothingness to be-ing, the perspective remains temporal. This is pure imagination, St. Thomas argued; for, passage or transition requires positive terms, but nothingness is not a positive term from which something could pass into be-ing. In order to understand creation, one must conceive it not as a mutation, not even perhaps as a production, which has temporal connotations, but as the absolute or transcendental causality of *esse*:

> Creation, insofar as its effect is concerned, implies two things. First, that the creature was not made from any preexisting matter; and, in this way, creation differs altogether from every kind of change or motion which always involves a substrate . . . Secondly, creation implies in the creature be-ing after non-being; not, however, according to any temporal succession, in the sense that a creature

St. Bonaventure: On the Eternity of the World (Milwaukee: Marquette University Press, 1964).

[20] Alexander of Hales, *Summa Theologica,* T. I, P. I, Tr. VI, qu. 2, c. 2 (ed. Quaracchi, I, p. 365): Si quaeritur quare non fuit mundus ab aeterno, cum voluntas sit causa proxima et sufficiens, et ab aeterno voluit mundum fieri; respondeo quod in intellectu creaturae est ens ab alio de nihilo, et omne ens de nihilo habet esse post non esse; unde in intellectu creaturae est aliquid quod repugnat aeternitati. – The editors add (Ibid., p. XXXIX): Tam profunde hanc quaestionem tractat Alexander de Hales ut argumentis suis fere nihil addere deinceps potuerint S. Bonaventura et Matthaeus Aquasparta.

[21] In II *Sent.*, I, 1, 5 (ed. Mandonnet, II, p. 33): Quae omnes [rationes] a philosophis tenentibus aeternitatem mundi positae sunt et solutae; et ideo potius in derisionem quam in confirmationem fidei vertuntur. – St. Thomas relied on Maimonides for the views of the philosophers, cf. L. Kendzierski, "Maimonides' Interpretation of the 8th Book of Aristotle's Physics," in *New Scholasticism,* XXX (1956), 37–48. The source of Maimonides' criticism of the Asharites was Averroes, *Kashf an Manahij al-Adillah,* p. 26. Cf. M. Fakhry, "The Antinomy of the Eternity of the World in Averroes, Maimonides, and Aquinas," in *Museon,* LXVI (1953), 139–55.

at one moment was not, then came into being, but according to a priority of nature or essence, in the sense that, left to itself, a creature would lapse into nothingness, since it has existence only through the Creator. For, a being has, of itself, what pertains to it by nature before what accrues to it from extrinsic causes . . . And according to these two considerations, creation is said to be from nothing. First of all, because the negation of any preexisting subject bars any material cause out of which the creature would be made; and, secondly, because creation implies a certain nothingness which the creature, by nature, is, since, by essence, the creature is non-existent. And if these two aspects are all that creation essentially implies, then reason can establish creation and, in that sense, the philosophers affirmed creation. If, however, one maintained that a third element is included in the notion of creation, namely a temporal succession from non-being to be-ing, then creation cannot be established by reason and, in that sense, it was not admitted by philosophers, although believers hold it on faith.[22]

We are faced here with an historical text of crucial importance for metaphysics. Basically, this view of creation is dependent on Avicenna's refutation of the arguments of the Mutakallimum for a temporal origin of the world. Because of his theory of cognitional emanations, Avicenna had been forced to defend both the necessity and the eternity of the creative flux against the Mutakallimum. However, he was able to retain what, in his eyes, was the genuine notion of creation by distinguishing two orders of causality: the metaphysical, which is the source of existence (*principium essendi*) and the physical, which is the origin of changes (*principium motus*). In making this momentous distinction, Avicenna seems to have been conscious of going beyond Aristotelian causality defined as "that from which change first begins."[23] Creation is needed to explain *esse*. Avicenna was neither listening

[22] In II *Sent.*, I, 1, 2, c (ed. Mandonnet, II, p. 18): Sciendum est autem quod ad rationem creationis pertinet duo. Primum est ut nihil praesupponat in re quae creari dicitur: unde in hoc ab aliis mutationibus differt, quia.... in mutationibus praesupponitus subjectum.... Secundum est, ut in re quae creari dicitur, prius sit non esse quam esse; non quidem prioritate temporis vel durationis, ut prius non fuerit et postmodum sit; sed prioritate naturae, ita quod res creata si sibi relinquatur, consequatur non esse, cum esse non habeat nisi ex influentia causae superioris. Prius enim unicuique inest naturaliter quod non ex alio habet, quam quod ab alio habet... Et secundum ista duo creatio dupliciter dicitur esse ex nihilo. Tum ita quod negatio neget ordinem creationis importatae per hanc praepositionem "ex" ad aliquid praeexistens, ut dicatur esse ex nihilo, quia non ex aliquo praeexistenti; et hoc quantum ad primum. Tum ita quod remaneat ordo creationis ad nihil praeexistens, natura affirmatus; ut dicatur creatio esse ex nihilo, quia res creata naturaliter prius habet non esse quam esse; et si haec duo sufficiant ad rationem creationis, sic creatio potest demonstrari, et sic philosophi creationem posuerunt. Si autem accipiamus tertium oportere ad rationem creationis, ut scilicet etiam duratione res creata prius non esse quam esse habeat, ut dicatur esse ex nihilo quia est tempore post nihil, sic creatio demonstrari non potest, nec a philosophis conceditur, sed per fidem supponitur.

[23] Avicenna, *Meta.*, Tr. VI, cc. 1–2, foll. 91ra–93rb.

to the *Loquentes nostrae legis,* as Averroes later accused him of doing, nor interpreting Aristotle, which for Averroes was the only philosophical role possible, but providing the elements of a metaphysics of creation. In his system, creation is the eternal source of existence for possibles or essences which, of themselves, tend to non-existence. St. Thomas, who always sought the *intentio profundior* of philosophical systems, was able to appreciate, far more than Averroes, what Avicenna was driving at. There is no doubt that, in his controversy about the eternity of the world, St. Thomas relied greatly, not to say principally, on Avicenna. As A. Pegis wrote:

> I propose to argue that the idea of creation was for St. Thomas a philosophical one, and that in its philosophical meaning it was not of Christian origin. It came, as he thought, not from the Greeks but from a highly gifted and original Arab commentator on Aristotle, Avicenna. Philosophical in origin, the idea of creation gave full and authentic expression to the Christian teaching that God was the author and maker of the universe . . . When St. Thomas read in Avicenna that *omne creatum est ens post non ens posterioritate essentiae* and that *creatio* is defined as *dare rei esse post non esse absolute* [cf. Avicenna, *Meta.*, Tr. VI, c. 2, fol. 92ra], he surely had in his hands the key to the philosophical doctrine of creation and through Avicenna he could isolate the philosophical meaning of the doctrine from the more complex theological idea of a creation in time. More than this, when St. Thomas looked at Avicenna's teaching on creation in the light of his own metaphysics of *esse,* all the questions about creation that other theologians had resolved in terms of duration he was able to see as ontological moments in the structure of the creature as such. The creature is *ens post non ens* by essence and absolutely.[24]

Historically, nothing contributed more to bringing out the existential, and so the transcendental, character of creation than the controversy over the eternity of the universe. First, the controversy clarified the metaphysical notion of creation by freeing it of all temporal connotations; secondly, the controversy related creation to the existence of the universe.

(1) By freeing the notion of creation from implications of coming-to-be after not being, and so of temporal inception, the controversy served to bring out a metaphysical definition of creation as simply a relation of dependence for existence, whether eternal or not. The possibility, philosophically speaking, of an eternal universe shows the inadequacy for metaphysics of such definitions of creation as the production of the whole entity of things from

[24] A. Pegis, "St. Thomas and the Origin of the Idea of Creation," in *Philosophy and the Modern Mind,* ed. Francis Canfield (Detroit: Sacred Heart Seminary, 1961), 49–50, 60. This article was written to illustrate the work of a theologian who carefully distinguishes philosophical from theological issues, and selects properly philosophical considerations for philosophical conclusions.

nothing. To produce is to bring something into existence which was not there before. In an eternal universe, there is no question of inception, beginning, or coming-to-be. The notion of creation has accordingly to be purified of connotations of production, bringing into being, transition from non-being to being, etc. Such terms are ineradicably temporal. They are used in theology, which accepts on faith the absolute beginning of the universe, but they are unfit for metaphysics which must abstract for time or eternity.[25]

Creation is a relative term: it can mean the creative act or its effect. In the first sense, creation can be called the emanation or flow of *esse*. Emanation is a Neo-Platonic term used by Dionysius and Avicenna to describe creation, and St. Thomas did not scruple to borrow it from them.[26] Emanation retains the idea of causal dependence or causal origin, but abstracts from temporal inception. In Avicenna's system, emanation is necessarily eternal; in Dionysius, it is an act of bounty. The term is neutral enough to serve in a metaphysical definition of creation. In the second sense, as term or effect of the creative act, creation is a real relation of total dependence of the creature for its *esse*. Or, to be more exact, it is the creature itself in its dependence on the Creator for its existence.[27]

Avicenna introduced that existential dimension of creation in metaphysics through his distinction between divine agents and natural agents. All divine agents cause *esse*.[28] In that perspective, creation is truly transphysical or metaphysical. That transvaluation makes Avicenna's and Averroes' world-views incommensurable. The ontological dependence of eternal processes of

[25] Cf. *De Pot.*, 3, 2. Time is coterminous with the universe. We cannot speak of a time before creation except in terms of imaginary time. But, in an eternal universe, there would be no first moment of time; whereas, in a temporal universe, it would be possible, in principle, to count back to a first moment. J. V. Peach, "The Age of the Universe," in *Heythrop Journal*, III (1962), 112–16, shows that there is nothing in the cosmic scales of astrophysics to make the eternity of the universe a necessity, nor to date the moment of creation.

[26] Cf. In II *Sent.*, Prol. (ed. Mandonnet, II, p. 1.); In II *Sent.*, 18, 1, 2 (ibid., p. 451); *Contra Gentiles*, II, 15; I, 45, 1.

[27] In II *Sent.*, I, 1, 2, and 4: Creatio . . . est quaedam habitudo habentis esse ab alio. I, 45, 3: Deus, creando, producit res sine motu. Subtracto autem motu ab actione et passione, nihil remanet nisi relatio. Unde relinquitur quod creatio in creatura non sit nisi relatio quaedam ad creatorem, ut ad principium sui esse.

[28] St. Thomas frequently gives credit to Avicenna for making the distinction between the two orders of causality. Cf. In I *Sent.*, VII, 1, 1, ad 3 (ed. Mandonnet, I, p. 178): Unde Avicenna, tract. X Metaph., cap. IX, dicit quod agens divinum differt a naturali: agens enim naturale est causa motus, sed agens divinum est dans esse totum, sicut creator mundi. In I *Sent.*, IX, 2, 2, praeterea (ibid., p. 252): Avicenna, tract. VI Metaph., cap. II, distinguit duplex agens: quoddam quod est causa fiendi tantum . . ., quoddam quod est principium essendi. In II *Sent.*, I, 2, 1, ad 1 (ibid., II, p. 46); XV, 1, 2, c (ibid., II, p. 372).

generation and corruption on a Final Cause which defined creation for Averroes, is transformed by Avicenna into an ontological dependence of finite existents for their *esse* on a source of *esse*.

(2) The second point that creation reveals is that *esse* is the fundamental reality which is limited and diversified by essence and individuation.[29] Essence, in finite beings, is reduced to a limiting role within *esse*. Although Avicenna failed to see the relation of essence to *esse* as that of limiting co-principle to its existential act; nevertheless, his conception of essence as, of itself, indifferent either to being or to non-being helped St. Thomas to regard *esse* as act of existing. Thus, in the *Sentences*, the *De Potentia*, the *De Aeternitate Mundi*, the *De Substantiis Separatis*, St. Thomas repeatedly states that essence, of itself, is nothing.[30] It is by *esse* that an essence is; and so it is by *esse* that an essence is an essence. What has a creature that it can call its own? Its identity: its self, its nature, its powers; but never its *esse*. Yet, without *esse*, what are its self, its nature, its powers? Nothing! As the *De Potentia* puts it: before it has *esse*, a creature is nothing, except perhaps in the creator's thought where it is not a creature but the creative essence.[31]

This is the conclusion to which this study has been pointing, namely, that essence, nature, properties, individuation constitute the morphology of *esse*.

[29] Cf. E. Gilson, *The Christian Philosophy of St. Thomas Aquinas* (New York: Random House, 1956), p. 142: "Now the act of being is like *the very stuff* from which things are made."

[30] St. Thomas refers to Avicenna on this point, cf. In II *Sent.*, I, 1, 2, c (ed. Mandonnet, II, p. 18, quoted above in footnote 22), and I, 1, 5, ad 2 (ibid., p. 38): Avicenna in sua Metaph. dicit ... unicuique enim est prius secundum naturam illud quod est ei ex se, quam id quod est ei ab alio. Quaelibet autem res praeter Deum habet esse ab alio. Ergo oportet quod secundum naturam suam esset non ens nisi haberet esse a Deo. *De Potentia*, 2, 1, obj. 11: Secundum Avicennam, quando res aliqua habet aliquid tantum ab altero, ei secundum se consideratae attribuitur oppositum ejus, sicut aer qui non habet lumen nisi ab alio, secundum se consideratus est tenebrosus, et per hunc modum omnes creaturae, quae habent ab alio esse, veritatem et necessitatem, secundum se consideratae, sunt non entes, falsae et impossibiles. Ibid., 4, 2, ad 14: Creatura, ex hoc quod est ex nihilo, habet tenebras possibilitatis et imperfectionis. *De Aeternitate Mundi*, n. 7 (ed. Perrier, p. 57): Prius enim naturaliter inest unicuique quod convenit sibi in se quam quod solum ex alio habet. Esse non habet creatura nisi ab alio; sibi enim relicta in se considerata nihil est; unde prius naturaliter est sibi nihilum quam esse. *De Substantiis Separatis*, VI, n. 45 (ed. Perrier, p. 151): Si igitur per hoc quod dico non ens removeatur solum esse in actu, ipsa forma secundum se considerata est non ens, sed esse participans. Si autem non ens removeat non solum ipsum esse in actu, sed etiam actum seu formam per quam aliquid participat esse, sic materia est non ens, forma vero subsistens non est non ens sed est actus qui est forma participativa ultimi actus qui est esse.

[31] *De Pot.*, 3, 5, ad 2: Ex hoc ipso quod quidditati esse attribuitur non solum esse sed et ipsa quidditas creari dicitur, quia antequam esse habeat nihil est nisi forte in intellectu creantis, ubi non est creatura, sed creatrix essentia.

Essence is the intrinsic determination and limitation, in an existent, of its *esse*. *Esse* is act, absolutely.

In a metaphysics of *esse commune*, *esse* and absolute act are mutually involved. They reveal themselves by implication, by the evidence which each sheds on the other, until the notion of *esse* as intensive act – the source of all being and perfection – is reached.

For all a being's perfection is due to its *esse*; a man, for instance, would be no better for all his wisdom unless by it he were wise. And the same holds for every other perfection. The measure of a being's perfection is its degree of *esse*: a being is said to be more or less perfect in proportion to the contraction of its *esse* to a greater or lesser mode of excellence. Hence if there were a being to which the whole perfection of *esse* belonged, no excellence possessed by anything would be lacking to it. Now a being which is its *esse* possesses *esse* according to its whole perfection.[32]

From that description of *esse* as intensive act, we can conclude that *esse* is the one transcendental perfection which subsists in itself, and that all the other transcendental perfections – unity, truth, goodness, beauty – are but aspects of *esse intensivum* from our way of conceiving and understanding being.

We now come to the moment of truth. So far, we have been speaking of finite or predicamental essence as the intrinsic principle of the limitation of *esse*. But is essence, of itself, necessarily a limit or principle of limitation? Is not essence an absolute perfection? For *esse*, if infinite, does not subsist without essence, but both essence and *esse* transcend their reciprocal limitations to become indentified in subsisting *Esse*. The question raises a delicate problem for Thomists. B. Montagnes has described two trends in recent Thomistic studies in metaphysics which flow from different perspectives: (1) in a perspective of *esse commune*, the *perfectio essendi* or *virtus essendi* is identified with *esse* as *actus essendi*, and the tendency is to regard essence as limitation; (2) in a perspective of *ens in communi*, the *perfectio essendi* comprehends the whole being, i.e. the subject, its essence, and its existential act.[33] For Montagnes, the second perspective retains the transcendental char-

[32] *Contra Gentiles*, I, 28: Omnis enim nobilitas cujuscumque rei est sibi secundum suum esse; nulla enim nobilitas esset homini ex sua sapientia, nisi per eam sapiens esset, et sic de aliis. Sic ergo secundum modum quo res habent esse, est suus modus in nobilitate. Nam res, secundum quod suum esse contrahitur ad aliquem specialem modum nobilitatis, majorem vel minorem, dicitur esse secundum hoc nobilior vel minus nobilis. Igitur si aliquid est cui competit tota virtus essendi, ei nulla virtus nobilitatis deesse potest quae alicui rei conveniat. Sed rei, quae est suum esse, competit esse secundum totam essendi potestatem.

[33] Cf. *La doctrine de l'analogie de l'être d'après saint Thomas d'Aquin*, pp. 159–68. Reviewing C. Fabro's *Participation et causalité selon S. Thomas d'Aquin* in *Bulletin*

acter of essence and accounts for the fact that St. Thomas always refused to say that the Creator has no essence, although he was fully aware of Avicenna's position on this point. J.-H. Nicolas points to the need of distinguishing between essence as principle of limitation and essence as correlative of *esse*. It is only in finite existents, where essence is really distinct from the act of existing, that essence is a limitation of *esse*. Where *esse* subsists in itself, in the Creator, essence is identified with His pure act of existence.[34]

In either case, whether essence is considered as transcendental perfection or whether it is considered as the principle of limitation of *esse* in creatures, the primacy of *esse* as *actus essendi* is unaffected. Essence is within *esse* – either by identity or by actuation. If we cannot avoid speaking of the essence of God, the reason may be our abstractive mode of conceiving. It is we who distinguish various perfections in God, and then we are obliged to correct this imperfection of our knowledge by identifying those various aspects with the *esse* of God. For all structures, whether metaphysical or noetic, are limitations either in things or in knowledge. We know that *esse intensivum* precontains eminently all the perfections which are diffracted throughout creation. And so, in order to be complete, the inversion of metaphysics requires that we regard *esse* not only as primary but as originative. And this means that finality, exemplarity, efficiency all flow from *esse intensivum*.[35]

3. The Immediacy of Creation

A genetic metaphysics proceeds from effect to cause. It starts from exist-

thomiste, XI (1960-1962), 20–21, and E. Gilson's *Introduction à la philosophie chrétienne* in *Revue des sciences philosophiques et théologiques*, XLV (1961), 719–20, Montagnes criticizes both authors for their tendency to speak of essence as principle of limitation of the absolute perfection of *actus essendi*. In the review of Fabro's work, Montagnes indicates the awkwardness of considering a finite being as composed of a predicamental element (essence) and a transcendental element (*actus essendi*). That objection seems to be a matter of perspectives. All that Fabro means is that being is either pure act of existing or is composed of potency (essence) and act (*esse*).

[34] J.-H. Nicolas, review of B. Montagnes' *L'analogie de l'être*, in *Revue thomiste*, LXIII (1963), 428–29. See also Nicolas' review of Fabro's *Participation et causalité* in *Revue thomiste*, LXII (1962), 277.

[35] Cf. *Contra Gentiles*, I, 28: Primum esse quod est fons essendi.... I, 25, 3, c: Esse divinum, super quod ratio divinae potentiae fundatur, est esse infinitum, non limitatum ad aliquod genus entis, sed praehabens in se totius esse perfectionem. *Contra Gentiles*, I, 84: In eo quod repugnat rationi entis inquantum hujusmodi, non potest salvari similitudo primi esse, scilicet divini, quod est fons essendi. R. Henle, *Saint Thomas and Platonism*, pp. 358–61, shows the unsuitability of any theory of divine ideas in Thomism to express ontological exemplarity, since the strict medium of knowledge is rather one idea, i.e. the divine essence. We are saying that Henle's correction should be extended, beyond essence, to *esse intensivum*.

ence as a datum which is not ultimately explained by natural processes of transformation and transmutation. To explain the existence of the universe, as Avicenna saw, there is need of a creative cause. With Avicenna, however, the account of creation became complicated with the Neo-Platonic problem of the diremption from the One to the many, so that creation became mediated by ten Intelligences and spheres down which it passed.[36] Christian theologians met the problem of mediation in creation in commenting on the *Book of Sentences* of Peter Lombard. While rejecting the fact of mediating causes in creation as contrary to the faith, they argued about the abstract possibility of instrumentality in creation. In Books II and IV of his commentary on the *Sentences*, St. Thomas questioned the arguments advanced by his contemporaries denying the possibility of a ministerial role by angels in creation. One objection, reported by St. Thomas, was to the effect that a creative agent must be by identity his action, i.e. must act *se toto*, and that no creature can realize this condition.[37] To that, St. Thomas answered that angels, being altogether immaterial or pure forms, could operate by essence or *se totis*. The answer is all the more surprising since St. Thomas had previously asserted that no being could be identically its operation unless it were identically its *esse*.[38]

The answer of St. Thomas does not go beyond Avicenna's position. Avicenna had distinguished two modes of operation: one by essence and the other by powers or *habitudines*. Sometimes he spoke more vaguely of agents who act *per se* and others who act by participation; but there is no doubt that the first mode of operation was that of divine agents or creators, and the second mode that of natural agents.

Back of the problem, in fact, is the whole story of contact [ἁφή] required to explain motion. Aristotle defined motion as the "actuation of the mobile as mobile; and the cause of motion is contact, so that the mover is also reacted upon." [39] He was therefore led to conclude that "the necessity that there should be motion continuously demands that there should be a first

[36] The axiom: Ab uno primo et simplici non procedit nisi unum, seemed to call for intermediaries to explain multiplicity. In I *Sent.*, II, 1, 1, obj. 1, quotes the axiom and refers to Aristotle. In Aristotle, the principle seems to apply to the uniformity of the motion of the first heaven. Cf. L.-B. Geiger, *La participation dans la philosophie de s. Thomas d'Aquin* (Paris: J. Vrin, 1942), pp. 100–101.

[37] In IV *Sent.*, V, 1, 3, sed contra 2 (ed. Moos, p. 207, n. 42), and ad tertiam questionem (ibid., p. 209, n. 57).

[38] Cf. in I *Sent.*, III, 4, 2, praeterea (ed. Mandonnet, I, p. 115): Ut dicit Avicenna, lib. *De Intellig.*, cap. 1, omne secundum agens agit inquantum participat aliquid; et ita agit per aliquid additum essentiae. . . . Also, ibid., VII, 1, 1, ad 1; XIV, 1, 1, ad 3; In II *Sent.*, XVII, 1, 2, ad 6.

[39] *Phys.*, III, 2, 202 a 3–4.

mover that is unmoved even accidentally." [40] This necessity led Aristotle to conceive the unmoved mover as acting by way of finality and not by way of efficient causality.[41] He accordingly distinguished two forms of contact: physical or extended contact and spiritual or intentional moving.[42] The only influence which higher causes exercise on lower causes is by way of finality. Aristotle did not consider causal participation in the order of efficiency.[43] But in Neo-Platonism, causal participation gave rise to various theories concerning the powers or *habitudines* by which, in the hierarchy of mediating causes, a lower cause shares in the perfection of a higher cause in order to produce so-called universal effects. These dispositions were regarded as either essential or accidental. The distinction marked off the universal causes from the particular causes.[44] Avicenna reserved agent causality to the creators; natural causes only disposed matter, through qualitative changes, to receive forms from the Giver of Forms, the last of the Intelligences.[45] In general, these theories considered all pure forms as intermediaries in creation by reason of their immateriality.

The only solution was to lift the whole problem out of the order of change and coming-to-be and to transpose it into that of *esse:*

> It is only in order to cause something by way of motion that an instrument is ever employed; for to be a moved mover is the essence of an instrument. Creation, however, is not a motion.[46]

[40] *Phys.*, VIII, 6, 259 b 23.

[41] *Meta.*, XII, 7, 1072 a 20 – b 14.

[42] *De Gen. et Corr.*, I, 6, 323 a 24 – 33. On motion by inspiring desire and love.

[43] Cf. A. Hayen, *L'intentionel selon saint Thomas* (Paris: Desclée de Brouwer, 1954), pp. 99–100.

[44] Cf. I–II, 50, 6, c: Quidam posuerunt in angelis non esse habitus, sed quaecumque dicuntur de eis, essentialiter dicuntur. Unde Maximus, post praedicta verba, dicit: habitudines earum, atque virtutes quae in eis sunt, essentiales sunt, *propter immaterialitatem*. Et hoc etiam Simplicius dicit in Commento Praedicamentorum: sapientia quae est in anima, habitus est; quae autem in Intellectu, substantia. Omnia enim quae sunt divina, et per se sufficientia sunt et in seipsis existentia.

[45] St. Thomas regarded the Avicennian position as Platonic, e.g. In II *Sent.*, I, 1, 4, ad 4 and XVIII, 2, 3, c (ed. Mandonnet, II, pp. 27, 469). At stake was the Aristotelian theory of *fieri*. The Avicennian theory is the background of such articles of St. Thomas as *Utrum creatio admisceatur in operibus naturae* in I, 45, 8; *De Pot.*, 3, 8; *De Virt. in comm.*, 11. St. Thomas took up this problem again in IV *Sent.*, I, 1, 4 (ed. Moos, p. 32, n. 124) which marks a date in the Saint's theory of instrumentality, cf. H. Dondaine, "A propos d'Avicenne et de saint Thomas; de la causalité dispositive à la causalité instrumentale," in *Revue thomiste*, LI (1951), 441–53.

[46] *Contra Gentiles*, II, 21: Instrumentum autem nunquam adhibetur ad causandum aliquid nisi per viam motus; est enim ratio instrumenti quod sit movens motum. Creatio autem non est motus. The distinction of the two orders, that of transformations and that of *esse*, and the elevation of creation to a truly transcendental and existential level is worked out in the *Expositio super Librum de Causis*, 20 (ed. Saf-

By the time of the Contra Gentiles, II, 21, and of the *De Potentia,* 3, 4, St. Thomas had found the answer that he was groping for when writing the *Sentences.* The condition for essential operation, and so for total causality, is not just immateriality, as the *Sentences* suggested, but pure act of existence. In order to act *se toto,* a being must be subsisting *esse*; for (1) nothing operates unless it exists, and (2) its operation must be identically its *esse.* To be total, a cause must be *Ipsum Esse Subsistens* whose proper effect is precisely *esse.*[47] The solution provides the last link between creation and *esse* as *actus essendi.*

In a way, the solution is nothing more than the traditional doctrine of potency and act. It says no more than that act is, absolutely speaking, prior to motion and change. It repeats the Aristotelian axiom that, in the order of change, potency precedes act, but that, in the order of being, act is absolutely prior as initiator of motion. However, in Aristotle, the principles are invoked to explain change and processes; whereas, in creationism, they are geared to existence. The medieval controversies about creation, which we have mentioned, served to mark off the two orders of motion and *esse.* It is this distinction which makes possible an existentialist account of the universe as opposed to an essentialist explanation. The controversy over the possibility of an eternal universe served to interiorize potency and act as components of the existential structure of creatures. It thereby made it possible to consider creation, not from an essentialist viewpoint as the production of the whole entity of things from nothing, but from an existentialist viewpoint as the emanation of *esse.* Likewise, the second controversy about intermediaries in creation served to link total causality with pure act of existence. This conclusion makes it possible to pass from such expressions as that God, acting by essence, produces the whole substance of things, to existential formulae such as that God, in giving *esse,* thereby produces what receives

frey, pp. 110–11): Ex praemissis concludit *impermixtionem causae primae* ad res alias. Et hujus conclusionis intellectus plenus haberi potest si accipiamus verba quae sunt in commento Procli.... (1) [Proclus] vocat autem "habitudinem" aliquam dispositionem per quam agens coaptatur seu proportionatur patienti seu recipienti.... Et sic tale agens secundum diversas dispositiones commiscetur rebus in quas agit secundum quamdam coaptationem ad ea; (2) sed causa prima agit per esse suum, ut probatum est. Unde non agit per aliquam habitudinem vel dispositionem superadditam per quam coaptetur et commisceatur rebus.

[47] *Contra Gentiles,* II, 22: Omnis virtus perfecta ad ea omnia se extendit ad quae suus per se et proprius effectus se extendere potest; sicut aedificativa ad omnia se extendit, si perfecta sit, quae possunt rationem habere domus. Virtus autem divina est per se causa essendi, et ejus effectus proprius est esse... *Quodlibet,* 12, 5, 5, c (ed. Mandonnet, p. 430): Quia esse est complementum omnium, inde est quod proprius effectus Dei est esse.

esse. From an essentialist *Deus se toto agit rem secundum totam ejus substantiam* we pass to an existentialist *Deus, simul dans esse, producit id quod esse recipit*. The existentialist expression is the metaphysical concept of creation.

CHAPTER V

THE TWO ORDERS OF CAUSALITY

In metaphysics, creation may best be described as the emanation of *esse*. Emanation retains something of the idea of source or causal origin without necessarily implying temporal inception. The priority of the source need be only ontological: that is, priority with regard to dynamic participation and causal dependence. Even in the supposition of an everlasting universe, creation would still connote derivation; whereas preservation in existence would connote causal participation continuing in time.

Needless to say that it is we, immersed in time, who denominate the Creator's action in different ways according to the effects that we distinguish in the universe. Divine action is creation, if the effect is participation in existence; it is conservation, if the effect is a being's enduring existence; it is divine motion, if the effect is a being's activity. One and the same act, which is the Creator's subsisting *esse*, is variously denominated from its effects. The denominations express the universe's total dependence on the Creator for existence and operation. What needs stressing is that creation, being transcendent causality, is supra-temporal. Whether the universe be finite or infinite in duration, creation includes time together with matter and motion and space. Far from being invariant frameworks, independent of the phenomena occurring in them, time and space are functions of matter and motion. Time and space are concreated with matter and motion. It is evidently not enough, so far as time is concerned, to push back indefinitely into an imaginary past the "moment of creation." That would prove only that the universe could have started earlier, not that it need never have started. What needs stressing is that creation is not a transition or coming-to-be. It does not belong to the order of time and duration, but to the order of *esse*. And the basic reason why *esse* cannot originate through process or transformation, is that this act is absolutely first in the order of being: it presupposes nothing, whereas all other actuations depend on it.[1]

[1] *Qu. Disp. de Anima,* 6, and 2: Ipsum esse est actus .... qui participabilis est ab

This radical difference between creation and processes sets off the two orders of causality: the transcendental and the predicamental. It is a fundamental point of Aristotelian causality that operation does not change the agent as such.[2] This point holds as well for predicamental activity as for creation. But in predicamental causality action mediates between the agent and the subject undergoing change. All the causing, though from the agent, is in the recipient. The same motion considered as *from* the agent is action, considered as *in* the recipient is passion. Through this mediation of action, predicamental agents are involved in a series of cosmic events which unfolds in time. The powers and determinations through which all secondary agents operate are the ground or foundation of the permanence of the predicamental relations which bind them into a cosmos. The interdependence of predicamental agents thus forms a network of mutual relations which constitutes the bond of the universe.[3]

But in creation there is no intermediary entity between the Creator and the universe. On the one hand, the creative fiat or imperium is the Creator's subsisting *esse*, eminently equivalent, by virtue of its pure actuality, to the powers and determinations by which secondary agents operate. On the other hand, creation is total and unconditioned causality: there is no mediating motion or *actio media*, in the absence of any potential, but absolute communication of *esse*. Actively considered, in the Creator, creation is the divine *esse*; passively considered, in the creature, creation is a pure relation of dependence:

> for, subtract motion from creation, and all that remains are the diverse Creator-creature relations.[4]

The difference between creation and physical processes leads to a startling

omnibus, ipsum autem nihil participat. *De Potentia,* 7, 2, ad 9: Nihil autem potest addi ad esse quod sit extraneum ab ipso, cum ab eo nihil sit extraneum nisi non-ens.

[2] *On the Soul,* III, 2, 426 a 4. Cf. J. McWilliams, "Action Does not Change the Agent," in *Philosophical Studies in Honor of the Very Reverend Ignatius Smith, O.P.*, ed. J. K. Ryan (Westminster, Md.: The Newman Press, 1952), pp. 209–21. This principle is true of changes which are strictly transformations or actuations of a potential; for there are processes, such as radiations, which involve quantitative losses, cf. P. Hoenen, *Cosmologia* (Romae: Aedes Pont. Universitatis Gregorianae, 1956), pp. 237–47.

[3] Cf. C. Kossel, "St. Thomas's Theory of the Causes of Relations," in *The Modern Schoolman,* XXV (1947-1948), 162–67.

[4] I, 45, 2, ad 2: Subtracto motu, non remaneant nisi diversae habitudines in creante et creato. *De Potentia,* 3, 3, c: Si sumatur active, creatio designat Dei actionem, quae est ejus essentia, cum relatione ad creaturam; quae non est relatio realis, sed secundum rationem tantum. Si autem passive accipiatur, cum creatio proprie loquendo non sit mutatio, non potest dici quod sit aliquid in genere passionis, sed est in genere relationis.

conclusion. Since passive creation is a relation of dependence of the creature on the Creator, it is ontologically posterior to the existence of the creature. In other words, the creature is not the term of creation, but its subject.[5] In a creationist perspective, the articulations are not: 1) God, 2) the coming-to-be of the universe, 3) the universe; but 1) God, 2) the universe, 3) its relation of dependence on God.[6] There is no coming-to-be mediating between the Creator and the universe. The universe exists, and it depends on the Creator. Its dependence is the condition of its existence, and so it is this dependence itself which is its creation. Passive creation is a real relation which is an accident in the creature since it presupposes the creature in existence.[7]

1. The Conservation of Beings in Esse

In creatures, then, the reality of creation is precisely their dependence for existence. But, if this relation is a supervening accident, does it not follow that beings are creatures only accidentally and not essentially?

The answer is that, although the relation of creation is accidental to finite beings, the dependence is total. For it is dependence in *esse*. The creation of

[5] *Contra Gentiles,* II, 18: Non enim est creatio mutatio, sed ipsa dependentia esse creati ad pricipium a quo instituitur, et sic est de genere relationis; unde nihil prohibet eam in creato esse sicut in subjecto. I, 45, 3, ad 3: Creationis, secundum quod significatur ut mutatio, creatura est terminus; sed secundum quod vere est relatio, creatura est ejus subjectum, et prius ea in esse, sicut subjectum accidente. *De Potentia,* 3, 3, ad 2: Relatio praedicta non est quoddam creatum, sed concreatum, sicut nec ens proprie loquendo, sed inhaerens.

[6] This point has been repeatedly stressed by A. D. Sertillanges, *La création, La somme théologique,* ed. Revue des jeunes (Paris: Desclée et Cie., 1927), pp. 245–46; "La notion de création," in *Revue thomiste,* XIII (1930), 47–57; *L'idée de création et ses retentissements en philosophie* (Paris: Aubier, 1945), pp. 43–61.

[7] We can speak of creative causality only by analogy with created activity. Predicamental cause-effect relations are real accidental relations. When St. Thomas calls passive creation a *relatio quaedam* which is an *accidens quoddam,* he is underlining its analogical character. The *relatio quaedam* is not what is usually called a transcendental relation in the scholastic tradition: cf. J. Anderson, "Creation as a Relation," in *New Scholasticism,* XXIV (1950), 280. However, there is only proportionality between, for instance, father-son relationship and Creator-creature relations. Action-passion relations, depending on the predicamental exercise of causality, are mutual relations; but Creator-creature relations are real only on the part of the dependent creature. The reason for this is the *impermixtio causae primae* explained in *Expositio in Librum de Causis,* Prop. XX (ed. Saffrey, pp. 110–11). Ontologically speaking, the Creator is totally unaffected by His creation. For He does not depend on any preexisting matter to be transformed, nor on intermediaries, nor on a good to be obtained, nor on an end to be achieved. In a word, the Creator is eminently free. However, He is not indifferent. He could not care more; for His creation is pure liberality: *ex amore divinae bonitatis est quod Deus eam communicare vult* (*De Potentia,* 3, 15, ad 14).

the universe is not an historical event which took place in the remote past, the effects of which are still around. Creation endures as conservation. The reason for this is simple, although it reaches to the roots of causality: effects depend on their causes according to the nature of the causality. If a cause is productive only of the coming-to-be of its effect, the production lasts only as long as the action of the cause continues: when the builders return home of an evening, the building comes to a stop; but if the cause is productive of the *esse* of its effect, the effect can exist only so long as the cause maintains it in existence. As creation is the source of *esse*, and through *esse* of all that exists, nothing can last without the Creator's actual communication of *esse*.

To illustrate, St. Thomas uses the example of light. According to the physics of his day, light was thought to be instantaneous. It was considered a quality permeating diaphanous bodies, such as air or anything translucent, under the continuous action of its source. St. Thomas uses a very striking expression to describe its relation to the illumined medium: *non habet radicem*. Light has no permanency in the air. Unlike a certain degree of heat, for example, which may remain in water for some time after the water has been removed from the fire, if a ray of light is cut from its source, it vanishes instantly.[8]

Light, accordingly, was said to be in the air by way of intention. This term expresses either the influence or the presence in a subject of some other *precisely as other*. Whether present in a flowing way, as in the use of an instrument by a principal cause, or more permanently by way of determinant, as in cognition or conation, an intention is never anything possessed by a subject as its own independently of the source and of its intrinsic reference to the other as other. An intention is a participation in an act or perfection which properly belongs to some other and is possessed dependently on that other.[9]

[8] I, 104, 1, c: Quia non habet radicem in aere, statim cessat lumen cessante actione solis.

[9] *De Potentia*, 5, 1, ad 6: Quaedam [formae] producuntur secundum perfectam rationem speciei et secundum perfectum esse in materia .... et hujusmodi formae remanent post actionem generantis usque as tempus corruptionis. Quaedam vero formae producuntur quidem secundum perfectam rationem speciei, non autem secundum perfectum esse in materia, sicut calor qui est in aqua calefacta ... Et hujusmodi formae possunt ad modicum remanere post actionem agentis; sed prohibentur diu permanere a contrario principio, quod est in materia. Quaedam vero producuntur in materia et secundum imperfectam speciem et secundum imperfectum esse, sicut lumen in aere a corpore lucido ... Unde sicut similitudo hominis non manet in speculo nisi quamdiu est oppositum homini, ita nec lumen in aere, nisi apud praesentiam corporis lucidi: hujusmodi intentiones dependent a formis naturalibus corporum per se, et non solum per accidens; et ideo esse eorum non manet cessante actione agentium.

Applied by analogy to *esse*, intentionality refers to the utter dependence of the creature for its existence on its Creator. This dependence is not a relation which flows from the principles of a being, but a dynamic relation of the *esse* of the creature to its source. Conservation in existence is, for finite beings, the very same relation as that which constitutes them creatures. Theirs is a permanently shared existence.[10] In the immutable actuality of His subsisting *esse*, the Creator attains directly and immediately whatever exists throughout the flux of time.

If the conservation of the universe in existence is in reality identical with its creation, both on the part of the divine act and of the relation of dependence in creatures, it would seem that conservation belongs to God alone. Yet, St. Thomas always attributed to secondary causes, especially to astral bodies, an instrumental role in the conservation of beings.[11] No mediation is possible in creation because there is no preexisting subject to creation; but, given beings in existence, God preserves some of them – those subject to decay and decomposition – through the agency of secondary causes. Everything created is created directly and immediately by God; but not everything conserved in existence is conserved by God uniquely and exclusively. There is what might be called a physical conservation of beings in existence which may be either direct or indirect. Physical conservation is direct, if the cause contributes positively in maintaining the effect in existence; indirect, if the cause counteracts or neutralizes disruptive influences. In either case, the conservation by secondary causes is purely a physical causality of the action-passion type. Two considerations, drawn from the parallel passages, confirm this view.[12]

(1) Although St. Thomas speaks freely of *dare esse, agere esse, producere esse*, there are times when he shows a studied preference for expressions such as *causare in esse, producere in esse, conservare in esse*.[13] For what

[10] *Contra Gentiles*, III, 65: Impressio agentis non remanet in effectu, cessante actione agentis, nisi vertatur in naturam effectus ... Esse autem non est natura vel essentia alicujus rei creatae ... Nulla igitur res remanere potest in esse, cessante operatione divina.

[11] Solar energy, for example, is indispensable for life on earth. Sun heat maintains conditions of life, for the most part, by preventing freezing and by keeping the water cycle going, Sunlight directly furnishes energy for photosynthesis. All living things ultimately derive energy from sunlight, plant cells directly, animal cells indirectly. But that is not exactly what St. Thomas had in mind when speaking of the universal causality of the celestial bodies. He attributed to them an overall intentional causality in the conservation of the universe, as we shall see in the following section.

[12] The main sources of this doctrine are: *Contra Gentiles*, III, 65 and 66; *De Potentia*, 5, 1 and 2; I, 104, 1 and 2; *Compendium Theologiae*, I, 131; *In Joann.*, 5, lect. 2.

[13] Cf. *Contra Gentiles*, III, 65 and 66 (Leonine ed., Appendix, 22 a 5): where *non*

is created and conserved is the existent, although it is through the act of be-ing that an existent is posited in the universe. And just as secondary agents bring new beings in existence, so can they preserve them in existence as causes of the endurance of things in existence.

(2) In the *De Potentia,* 5, 1, ad 6 (quoted on p. 61), St. Thomas explains in what sense secondary conserving causes can be called *causae essendi* by saying that they maintain their effects in an *esse in fieri.* The strangeness of the expression, as well as the examples of illumined atmosphere and mirror images, leave no doubt that the causality is physical. It is no different from the boiling of water on a fire over a period of time or the preservation of food by refrigeration.

With these distinctions, it is possible to co-ordinate the conserving causalities as follows: 1) on the transcendental level of *esse,* God conserves directly and immediately the existence of creatures by the same act by which He gives *esse;* [14] 2) on the predicamental plane of physical agency, created causes concur both positively and preventively in maintaining the physical integrity of corruptible natures. These operations of created causes are instrumental in keeping such natures in existence. But God always remains the principal cause of *esse,* for effects belong properly to the principal and not to the instrumental cause.[15]

In fine, no created agent is the direct and immediate cause of the *esse* of its effects. Created agents operate within a framework of existence preserved by the Creative Cause. Created agents attain *esse* indirectly through the transformations which they effect; for *fieri* terminates in *esse.* Any attempt

est causa nisi fieri artificiatorum is corrected to *non est causa artificiatorum nisi quantum ad fieri*; (ibid., 22 b 27): *causae secundae esse conservant* is corrected to *causae secundae in esse conservant.*

[14] Since the divine action is the divine *esse,* the immediacy is both of supposit and operation: cf. *Contra Gentiles,* III, 68; *De Potentia,* 3, 7, ad 7.

[15] The implication of direct conservation of beings in existence by God alone in I, 104, 1, ad 4: "conservatio rerum a Deo non est per aliquam novam actionem, sed per continuationem actionis qua dat esse," seems to be denied in the next article, I, 104, 2, sed contra: "per idem conservatur res, per quod habet esse. Sed Deus dat esse rebus mediantibus aliquibus causis mediis. Ergo res in esse conservat mediantibus aliquibus causis." M. J. Adler, "The Demonstration of God's Existence," in *The Thomist,* V (1943: The Maritain Volume), 206–208, sees a real discrepancy between the two texts. The ambiguity arises from the repeated statements by St. Thomas that directly conserving secondary causes are *causae essendi* of their effects, whereas they are only *causae permanendi in esse.* But, in article 1, St. Thomas is speaking of creation and conservation in the absolute and transcendental sense; in article 2, he is referring to *physical* processes by which new beings are brought into existence by secondary causes and to *physical* conservation of corruptible beings by natural causes. As causes of the endurance of their effects, these causes are loosely called *causae essendi.* It would be better to call them *causae permanendi in esse.*

to make *esse* exclusively the effect of God, except in creation proper, leads to an occasionalistic view of the production of *esse*: secondary causes would educe forms and God would give *esse*. Such a view vitiates the whole idea of instrumentality; for an instrument is used not to produce an effect proper to itself only, but rather to contribute by its proper activity to the precise effect intended by the principal agent.

2. *Universal causes*

Early in the *Sentences*, St. Thomas had pointed out that it was ignorance of creation that had led ancient philosophers to posit a number of autonomous first principles of the universe.[16] It is not clear how far St. Thomas ever realized how radically creation corrected this pluralism. The correction eliminates precisely the universal causality of the heavenly spheres.[17] When one recalls that even Copernicus accepted the medieval cosmology of crystaline spheres, ethereal stars, and uniform circular motions, one sees how humanly impossible it was for St. Thomas to make a clean break with such astronomical theories. Besides, the *Sancti* themselves, as well as the biblical imagery of the times, concurred in such views. They were the accepted facts. All his life, St. Thomas held in a general way the Aristotelian cosmic system of Prime Mover, meshing celestial spheres, and activated terrestrial processes. The classical expression of the all-pervading role of the heavens is: *Homo generat hominem, et sol.*[18] On that cryptic remark hangs the whole theory of universal causes. For Aristotle, it meant simply that the cycles of generation and corruption were maintained by the sun in its course round the ecliptic. The remark was pointed to the Academy. Plato had explained generations by saying that an individual man fathers another individual, but that both are men by participation in the Idea of Man, as they are living beings by participation in the Idea of Life and in the Idea of Being. Aristotle replied that his own real, observable causes dispensed with the Ideas.[19]

[16] In II *Sent.*, I, 1, 1, sol.

[17] Cf. J. de Finance, *Etre et agir*, p. 147: St. Thomas never fully realized how radical a transformation the metaphysics of creation imposes on the seried universe of the Greeks. A. Hayen, *La communication de l'être*, pp. 69–71: St. Thomas accepted without questioning the hierarchical structure of the Dionysian universe. C. Fabro, *Participation et causalité*, p. 426: One cannot escape the impression that Thomistic causality was developed along two antithetical lines: that of universal causes and that of creative cause.

[18] *Phys.*, II, 2, 194 b 13.

[19] Cf. W. D. Ross, *Aristotle's Physics* (Oxford: Clarendon Press, 1955), p. 511. T. Litt, *Les corps célestes dans l'univers de saint Thomas d'Aquin* (Louvain: Publications Universitaires, 1963), pp. 273–76.

With Neo-Platonism, however, the Aristotelian theory of heavenly influences became linked with another insight: that the cause of a species cannot be univocal, otherwise it would have to be cause of itself. In other words, no univocal cause can account for a form as such, but only for this form in this matter. The cause of any species must be a universal and equivocal cause.[20] The fortune of the heavenly bodies was made. Particular causes, accounting only for individual forms in matter, are causes of *fieri*; universal causes are *causae essendi* insofar as by their operation they tend to preserve the species. All these seried causes concur simultaneously in producing one and the same effect, e.g. a natural unit; but particular causes operate both in virtue of their own powers and in virtue of the influence of the celestial spheres, and these operate in virtue of their incorporeal movers, Intelligences or angels, and all in virtue of the Prime Mover.[21] In the cosmic hierarchy, all causes are moved except the first, and so all causes except the highest are instrumental. They combine to produce a same effect,

> so that no matter how intense a fire may be, it would not burn anything unless moved by astral bodies.[22]

In Avicennism, there was some excuse for such mediation, since the intermediaries were all creators; in Thomism, the theory of universal causes fits like an extra leg. If all this mediate causality were dispensed with, natural agents would still effect *esse hoc* as the term of *fieri*; but then they would be co-operating directly with the Creator in producing and preserving their effects in existence.

In order to produce a form as such, one would have to be able to give

[20] St. Thomas attributed this view to Avicenna, e.g. In I *Sent.*, 12, 1, 2, ad 2 (ed. Mandonnet, I, p. 292); In II *Sent.*, 18, 2, 1 (ibid., II, p. 460). Avicenna, *Meta.*, Tr. VI, 3, fol. 92va, and Tr. VI, 5, fol. 94va, called the *causa causati in sua specie* a *causa essentialis* or *natura universalis*; the *causa causati in suo individuo*, he called a *causa non essentiali ordine* and *natura particularis*. The distinction separated the creators from physical agents.

[21] *De Potentia*, 3, 7, c: Quia natura inferior agens non agit nisi mota, eo quod huiusmodi corpora inferiora sunt alterantia alterata; coelum autem est alterans non alteratum, et tamen non est movens nisi motum, *et hoc non cessat quousque perveniatur ad Deum*. Cf. R. O. Johann, "A Comment on Secondary Causality," in *The Modern Schoolman*, XXV (1947-1948), 24–25; T. Litt, *Les corps célestes dans l'univers de saint Thomas d'Aquin*, pp. 166–73.

[22] I–II, 109, 1, c: Quantumcumque ignis habeat calorem perfectum, non alteraret nisi per motionem corporis coelestis. *De Potentia*, 5, 8, ad 1: Virtus ignis semper est determinata ad calefaciendum, praesuppositis tamen causis prioribus quae ad actionem ignis requiruntur. Cf. B. Lonergan, "St. Thomas' Theory of Operation," in *Theological Studies*, III (1942), 386: What is affirmed with such confidence in the *De Potentia* and the *Prima Secundae* is only a probable opinion on the authority of Simplicius in the *De Coelo*, II, lect. 2, n. 13.

esse directly; for such a form would not be educed from any substrate, but given existence absolutely. And that is to create. In an existential metaphysics, there is only one universal cause: the Creator who is *Ipsum Esse Subsistens*. God is not only the first of a series of movers nor the initiator of a causal chain, but the immediate source of *esse* and operation in all beings:

> [Natural] agents operate from the outside: for, since they operate only by moving and changing things, they do so by acting *on* others. But God is immanent in things, since His action is creative. To create is to give *esse* to creatures. And as *esse* pervades everything, God who gives *esse* is immanent in all things.[23]

Esse is maintained by God directly. In *esse*, creatures develop according to the rhythm of their innate capacities. They constantly act and react on each other. All this activity belongs to the predicamental order of motion and change. Where premotion is necessary to explain the start of chains of action, such premotion belongs to the physical level of causality. Whether immediate or mediate, premotion is in the order of motion. It is part of the dynamic pattern of cosmic events, of the combinations and interferences of created causes, which form the history of the world. All created causes are on the same level. None is a universal cause.

3. *Universal instrumentality*

It is only indirectly, i.e. through transformations, that secondary agents attain *esse*. But given something to work on, they do bring new beings into existence. To express the juncture of the two orders of causality, the predicamental order of transformations and the transcendental order of *esse*, and to explain the role of secondary agents in producing new *esse's*, St. Thomas resorts to the analogy of instrumentality. Instrumentality is a correct technical term for the operational dependence of a lower cause on a higher one where the lower cause is insufficient, by itself, to account for an effect. The analogy, as an explanation of created activity, has recently been criticized on three scores. (1) It fails to account for the total dependence of creatures and their activity; for instruments have their own nature and use-

[23] *In Joann.*, 1, lect. 5, n. 133 (ed. Cai, p. 28): Alia agentia operantur ut extrinsecus existentia: cum enim non agent nisi movendo et alterando aliquo modo quantum ad ea quae sunt extrinseca rei, ut extrinseca operantur; Deus vero operatur in omnibus ut interius agens, quia agit creando. Creare autem est dare esse rei creatae. Cum ergo esse sit intimum cuilibet rei, Deus, qui operando dat esse, operatur in rebus ut intimus agens.

fulness which they need not hold from the principal cause. (2) The analogy stems from the Greek conception of a hierarchical universe comprising any number of universal causes but no Creator. It belongs to the order of moved-movers. (3) St. Thomas himself, in the *Summa Theologiae*, relegates the analogy to the background as a manuduction, and not as an explanation.[24]

The aim of these criticisms is to free St. Thomas' own insight from the often inadequate expression of his views. Often, in St. Thomas, the expression lags far behind the thought; for he often remains faithful to the verbal context which he is commenting while transcending his source. In the present instance, instrumentality would pertain to the all too exterior consideration of physical premotion; whereas participation would better convey the idea of the inwardness of the Creative Cause immanently operating in all reality.

However, without expecting more from the analogy than it is meant to yield, it may just be that it is the finest way of indicating the total dependence of creatures on the Creator in the production of new *esse's*. Instrumentality, as we have already mentioned, is one form of intentionality. It is a flowing intention from the principal cause through the subordinate cause to the effect: *ens ab alio ad aliud*. In plain words, it is the use of a means to achieve one's purpose. St. Thomas is quite free with the use of analogies. He calls secondary agents *quasi* instruments of God in the production of new *esse*. The *quasi* shows that he was conscious of the imperfection of the analogy to express the total dependence of the nature, powers, and operations of secondary agents on God. But instrumentality adds a precision to secondary causality. It indicates that secondary agents operate within a system in which they achieve something beyond their natural powers: they co-operate with God both in producing new *esse's* and in maintaining the order of the universe. In neither case are secondary causes principal causes. The order of the universe is a dynamic unity beyond the interactions of secondary agents, yet achieved through their activity. *Esse*, likewise, is beyond the power of secondary causes to produce directly. These two universal effects – world order and *esse* – belong to God as principal cause and to creatures as *quasi* instruments of God. Thus the activity of creatures, in the line of nature, is the proper although secondary cause of transformations; but only instrumental cause of *esse* and of universal order. Instrumentality is one form of dynamic participation.

Sometimes, even in the same context, St. Thomas says equivalently that

[24] C. Fabro, *Participation et causalité*, p. 506; J. de Finance, *Etre et agir*, pp. 226–29; A. Hayen, *La communication de l'être*, pp. 102–103; J. Rock, "St. Thomas on Divine Causality," in *Philosophical Studies*, V (1955), 21–44.

creatures are accidental or instrumental causes of *esse*.[25] An accidental cause is one which produces a certain effect indirectly, i.e. through something else which is directly intended; an instrumental cause produces an effect beyond what it is proportioned to by nature. Both expressions designate the indirect way in which secondary causes attain *esse*. (1) In all changes, what is directly effected is a new composite. Properly speaking, non-subsistent forms are not produced; they are coproduced through the change of one composite into another. These forms, as such, have only an accidental cause. The same holds for *esse*. This is the reason why no secondary cause can produce any form as such, i.e. none can give *esse directly*. (2) As for the second expression – instrumental cause – this brings out even more clearly that secondary agents are proper causes only of *fieri*, but that they participate in the production of *esse* by elevation to that order. Like the changes which it produces, efficiency is in the existential order. It is bound to issue in *esse*: no matter how badly a job is done, something *is* always accomplished.

Here we are at the heart of the metaphysical aspect of production: the point where predicamental causality fits into the framework of creation and conservation of *esse*. Creatures share in *esse*; their operations are existential acts, i.e. the motions and changes actuating things; the effects produced by these changes owe their existence to the created agents only as *quasi* instruments of the Creator. Not that secondary agents produce only transformations which God caps with *esse*; but that they attain *esse* only through *fieri*. All processes, and all agents involved in them, bathe in *esse*. For secondary causes, the order of *esse* is always a shared one: *esse* is the domain of the Creator.

This seems to be what the *quasi* instrumental character of secondary causality is meant to convey: the indirect and totally dependent way in which creatures attain *esse* which is the proper effect of *Ipsum Esse Subsistens*. In creation, *esse* is the medium through which all else is produced: matter, form, accidental determinations. In predicamental activity, form is the medium through which matter [a substrate] is actuated and the *esse* of the new existent is produced *per accidens*. In all the changes which secondary agents effect, *esse* is accounted for by the fact that agents participate in *esse*, that the changes which they operate are existential actuations, and that these issue in new existents who in turn participate in *esse*, and, above all, that God sustains agents, operations, and results in existence.

[25] Cf. *Contra Gentiles*, II, 21: *Amplius*: non prima causa, sed secunda et instrumentalis ... *Adhuc*: alia sunt causae essendi per accidens et non per se ...

SUMMARY AND CONCLUSION

In a lucid page, M. Fakhry has underscored the basic opposition between Aristotelianism and Neo-Platonism: the dualistic tendency of the theory of potency and act, on the one hand, and the monistic tendency of Neo-Platonic emanationism, on the other hand.[1] His concluding remarks are:

> I content myself here with noting the great difficulty to the vindication of creation *ex nihilo* which the Aristotelian distinction between Act and Potency raises. The notion of Act in Aristotle is dictated by the requirements of a dualist metaphysics and has meaning only in contradistinction to Potency. Aristotle is, therefore, perfectly consistent with himself when he carries this dualism to the extreme and sets Act and Potency, God and Matter up against each other, as two co-eternal principles. Creationism, on the other hand, can be rationally vindicated only in terms of a monistic metaphysics in which the initial distinction between Act and Potency does not as much as arise. How the Thomist doctrine of creation *ex nihilo* can be fitted into a metaphysics of Act-Potency is very difficult to see. A deeper reading of Aquinas would show that his doctrine of creation is of Neo-Platonic (notably Dionysian) extraction; an emanationist monism whose philosophical scaffolding is Aristotelian, and from which the determinist "sting" has been artfully removed. In fact, Aquinas speaks of creation as a process of "emanation" of things from the First Principle in the S. Theol., I, 45.[2]

To resolve the conflict arising from the opposite pressures of the two systems – the actuation of a primary uncaused potential and the emanation of essences from the One – it was necessary to rise to a higher level of consideration allowing synthesis. By transposing the problem into the existential order, always presumed but equally neglected by Aristotelianism and Neo-Platonism, St. Thomas situated creation on an absolutely transcendental plane of consideration.

[1] M. Fakhry, *Islamic Occasionalism and its Critique by Averroes and Aquinas*, p. 157.
[2] Ibid., p. 199.

(1) For, it is between *to be* or *not to be*, taken absolutely, that is, between *esse aut non esse simpliciter*, that there is an exhaustive division. So long as primacy is accorded to Form, or Οὐσία, or the One, there is always possibility of a wedge between being and non-being, namely, pure potency or matter.[3] And, in Plato and Aristotle and Neo-Platonism, matter is simply a datum. It is a pure facticity.

a) Averroes confirms this point. He rejected Avicenna's existential emanationism in order to return to Aristotle's substantialism. He, therefore, regarded creation as the source of actuations only (cf. pp. 33-34). Hylomorphism, for him, accounted for being. Since matter cannot exist without form, the cause of the composite of matter and form accounts for the presence of matter. Form gives existence; and the First Mover, as source of the eternal processes of generation and corruptions, is the source of existence. Admittedly, this view is not the biblical concept of creation; but, in metaphysics, the meaning of creation depends on one's understanding of being as being. It is doubtful whether one can establish metaphysically, in an essentialist perspective, a total causality of the whole entity of things *ex nihilo*. Historically speaking, at least, such a concept was never attained except through the influence of Revelation. It is a theological notion drawn from the revelation of God's causality (cf. pp. 35-36). In metaphysics, unless the existence of the universe is the issue, it is hard to see why, especially in the possibility of an eternal universe, it need be produced *ex nihilo*. The indestructibility of the substrate of physical processes – whether Aristotelian prime matter or the ultimate particles of atomists or the matter/energy complex of physics – guarantees the permanence of the universe. And, for the materialist, *to be* is to be a particle or a complex of particles. This view, however, cannot answer the Question of Being.[4]

b) The same facticity and metaphysical insufficiency can be charged to dialectical systems and process philosophies.[5] What is ὄν which is equally

[3] Cf. J. Gredt, *Elementa Philosophiae Aristotelico-Thomisticae*, I, p. 220, n. 261: Inter merum nihil et actum datur tertium quid: realis potentia ... i.e. ultima linea realitatis, pura potentia realis, qua actuatur seu evolvitur per formam et existentiam.

[4] Here is a curious remark of St. Thomas on this subject from the *Contra Gentiles*, II, 16: Esse autem est universalius quam moveri; sunt enim quaedam entia immobilia, ut etiam philosophi tradunt, ut lapides et hujusmodi. Oportet ergo quod supra causam quae non agit nisi movendo et transmutando, sit illa causa quae est primum essendi principium.

[5] Cf. *In Meta.*, XI, 9, nn. 2289–2308, especially 2305–2306: Unde relinquitur quod motus est actus existentis in potentia ... Et propter hoc difficile est accipere quid sit motus. Videtur enim quod aut necesse sit ponere motum in genere privationis [seu indigentiae], aut in genere potentiae, aut in genere actus simplicis et perfecti; quorum nullum contingit esse motum. Unde relinquitur quod motus sit id quod dictum est,

μὴ ὄν? Change, process, the *it is* and *it is not* locked in dynamic tension. Change replaces *esse*. *To be* means to be happening, to be occurring, to be an event. And since what exists are interactions and events, both dynamic rather than static terms, reality is fundamentally characterized by flux. A "thing" is but a still photograph, taken by the mind, of evolving reality. This view of reality is like the stream-of-consciousness theory extended to the universe. Useful in sciences delimited to measuring and correlating phenomena, such an abstraction cannot be the whole picture (cf. pp. 21-22). Change is characterized by intentionality, direction, and limit; for interactions and relations require terms. And what those terms *are* determines what happens and occurs when they interact.

(2) For St. Thomas, the biblical concept of total causality led to a deeper understanding of being insofar as it *is*. The medieval controversies about the eternity of the universe and the possibility of intermediaries in creation helped to focus the issue.

a) It was to Avicenna that St. Thomas turned for a metaphysical formulation of the theory of creation (cf. pp. 49-52). Avicenna had interpreted the Neo-Platonic emanationism as the flow of existence (cf. pp. 26-28). His cognitional emanationism, however, had forced him to regard creation as an eternal flux. Then, in order to answer the contention of the Kalam theologians that creation involves *esse post nihil* or temporal inception, he lifted the issue out of the order of motion and time by fusing "nothingness" with *esse*, not in a dialectical way, as in modern Existentialism, but as articulations of the metaphysical structure of creatures: essence and *esse*. For Avicenna, essences of themselves tend to nothingness unless sustained by the flow of existence emanating from the Necessary Being. In this existentialist perspective, creation is the giving of *esse* in an absolute way. This bold move of Avicenna helped St. Thomas to formulate his own theory of existence as *actus essendi* and *actualitas omnium perfectionum* (cf. p. 51).

b) The last point, the question of intermediaries in creation, completed the doctrine. In the *Sentences*, St. Thomas accepted the possibility of angels serving ministerially in creation. He knew Avicenna's position on this point, but he could not accept it; he had to advance other reasons for his own views. He argued that since angels are immaterial or pure acts, they can operate *se totis*. Being independent of material conditions, they do not require a substrate on which to operate. Rethinking this problem, St. Thomas came to see that in order to operate *se toto*, that is, to be by identity its operation, a being must be pure act, not in the order of Forms, but in the

scilicet actus; et quod non dicatur actus perfectus. Quod quidem difficile est videre, sed tamen contingens est esse, quia hoc posito, nullum sequitur inconveniens ...

order of *esse*. In other words, the Creator must be *Ipsum Esse Subsistens*: not only the Idea of the Good as in Platonism, nor the subsisting Thought of Aristotelianism, nor the One of Neo-Platonism, but subsisting *Esse*.

The conclusion reached is that (a) *esse* is the very perfection (ἐντελέχεια) and actuality (ἐνέργεια) of the Creator which is participated through creation, and (b) participation can be understood in terms of the limitation and diversification of *esse* by essence. Participation can be explained according to the fourfold causal pattern of Aristotle, but the pattern is now transposed from the order of motion, which it was originally meant to explain, to the order of *esse*.

From a genetic approach, *esse* emerges as *actus essendi* because *esse* is what the creature shares from the Creator. The viewpoint sets the intelligibility and analogy of *esse* within the framework of efficient causality as an effect-to-cause relationship: the *proportio plurium ad unum*. *Esse* and essence are seen to be related as act and potency, not by what *esse* and essence are in themselves, but by the relationship which an existent bears to its Creator.[6] The creature resembles God, its efficient and exemplary cause, to the extent that it *is*. The measure or degree of perfection of its *esse* is intrinsically determined by its essence as limiting principle.

> for nothing can be added to *esse* from the outside; hence *esse* cannot be determined by anything else as potency by act, but rather as act by potency.[7]

And that is the central insight to which a genetic approach in metaphysics leads. Existence is absolutely primary, and any determination of existence is a contraction of existence. A determination is both perfection and limitation: it is perfection, but only in a relative way, that is, in relation to other determinations in the hierarchy of beings; absolutely speaking, it is a contraction of *esse*. It falls short of the *virtus essendi*, the *perfectio essendi*, the *plenitudo essendi* which is fontal. Because it does, any determination of *esse* makes that supreme *diremption of being* from Subsisting *Esse* to the Many.

[6] Cf. texts quoted on p. 51.

[7] *De Potentia*, 7, 2, ad 9: Nihil autem potest addi ad esse quod sit extraneum ab ipso; unde non sic determinatur esse per aliud sicut potentia per actum, sed magis sicut actus per potentiam.

SELECTED BIBLIOGRAPHY

I. PRIMARY SOURCES

Patrologia Graeca, ed. J. P. Migne. 162 vols. Paris: Migne – Garnier Frères, 1857-66.

Patrologia Latina, ed. J. P. Migne. 221 vols. Paris: Migne – Garnier Frères, 1844-80.

St. Albert the Great. *Opera Omnia*, ed. August Borgnet. 36 vols. Paris: Vivès, 1890.

Alexander of Hales. *Summa Theologica* studio et cura PP. Collegii S. Bonaventurae edita. 4 vols. Quaracchi: Typographia Collegii S. Bonaventurae, 1924-48.

Alfarabi. *Fontes Quaestionum*, ed. Miguel Cruz Hernandes. *Archives* d'histoire doctrinale et littéraire du moyen âge, XVIII (1951), 306-23.

Algazel. *Algazel's Metaphysics*, ed. J. T. Muckle. Toronto: St. Michael's College, 1933.

Aristotle. *Metaphysics*, text with commentary, W. D. Ross. 2 vols. Oxford: Clarendon Press, 1958.

— *La métaphysique*, ed. J. Tricot. 2 vols. Paris: J. Vrin, 1948.

— *Physics*, text with commentary, W. D. Ross. Oxford: Clarendon Press, 1955.

Averroes. *Aristotelis metaphysicorum libri XIV cum Averrois Cordubensis in eosdem commentariis et Epitome*. Venetiis: apud Juntas, 1562.

— *Destructio Destructionum Philosophiae Algazelis* in the Latin Version of Calo Calonymos, ed. Beatrice H. Zedler. Milwaukee: Marquette University Press, 1961.

Avicenna. *Opera* in luce redacta ac nuper quantum ars niti potuit per Canonicos emendata. Venetiis, 1508. Reprint: Frankfurt am Main: Minerva, 1961.

St. Bonaventure. *Opera Omnia*. 10 vols. Quaracchi: Typographia Collegii S. Bonaventurae, 1882-1902.

Chalcidius. *Timaeus*, ed. J. H. Waszink. *Corpus Platonicorum Medii Aevi*, Vol. IV. Leiden: E. J. Brill, 1962.

Heidegger, Martin. *Being and Time*, tr. J. Macquarrie and E. Robinson. New York: Harper and Row, 1962.

— *Introduction to Metaphysics*, tr. Ralph Manheim. New York: Doubleday Anchor Books, 1961.

— *Kant and the Problem of Metaphysics*, tr. J. Churchill. Bloomington: Indiana University Press, 1962.

— *Vom Wesen des Grundes*. Halle: Max Niemeyer, 1929.

— *Brief über den Humanismus*. Bern: Francke, 1947.

— *Nietzsche*. Pfullingen: Neske, 1961.

Leibniz, G. W. *The Monadology and Other Philosophical Writings*, ed. Robert Latta. London: Oxford University Press, 1951.

Maimonides. *The Guide for the Perplexed*, tr. M. Friedlander. New York: Dover Publications, 1956.

Proclus. *The Elements of Theology*, tr. and ed. E. R. Dodds. Oxford: Clarendon Press, 1963.

Scotus, Joannes Duns. *Opera Omnia* studio et cura Commissionis Scotisticae edita, praeside C. Balic. 7 vols to date. Civitas Vaticana: Typis Polyglotis Vaticanis, 1950.

Siger de Brabant. *Quaestiones in metaphysicam,* ed. C. Graiff. Louvain: Institut Supérieur de Philosophie, 1948.

Suarez, Franciscus. *Disputationes Metaphysicae. Opera Omnia.* 28 vols. Paris: Vivès, 1856-77. Vols. XXV–XXVI.

St. Thomas Aquinas. *Opera Omnia,* ed. Leonia. 16 vols. to date. Rome: Typis R. Garroni, 1882.

— *Scriptum super Libros Sententiarum Magistri Petri Lombardi,* ed. P. Mandonnet and F. Moos. 4 vols. Paris: Lethielleux, 1929-47.

— *Le "De Ente et Essentia" de S. Thomas d'Aquin,* ed. M.-D. Roland-Gosselin. Paris: J. Vrin, 1948.

— *Expositio super Librum Boethii De Trinitate,* ed. Bruno Decker. Leiden: E. J. Brill, 1955.

— *Contra Gentiles,* texte de l'édition Léonine; introduction de A. Gauthier, traduction de R. Bernier, M. Corvez, F. Kerovanton. 2 vols. to date. Paris: Lethielleux, 1957-61.

— *Summa Theologiae* cura et studio Instituti Studiorum Medievalium Ottaviensis ad textum S. Pii V iussu confectum recognita. 4 vols. Ottawa: Impensis Studii Generalis O. Pr., 1941-45.

— *Somme Théologique, La création,* traduction et notes de A. D. Sertillanges. Paris: Desclée et Cie., 1927.

— *In Duodecim Libros Metaphysicorum Aristotelis Expositio,* ed. M.-R. Cathala and R. M. Spiazzi. Turin: Marietti, 1964.

— *In Octo Libros de Physico Auditu sive Physicorum Aristotelis Commentaria,* ed. F. Angeli and M. Pirotta. Naples: D'Auria, 1953.

— *Quaestiones Disputatae.* 2 vols. Vol. I, *De Veritate,* ed. R. Spiazzi; Vol. II, *De Potentia,* ed. M. Pession. Turin: Marietti, 1953.

— *Opuscula Omnia necnon Opera Minora.* Tomus I. Opuscula Philosophica, ed. J. Perrier. Paris: Lethielleux, 1949.

— *Opuscula Omnia* genuina quidem necnon spuria melioris notae debito ordine collecta cura et studio P. Mandonnet. 5 vols. Paris: Lethielleux, 1927.

— *Quaestiones Quodlibetales,* ed. P. Mandonnet. Paris: Lethielleux, 1926.

— *Tractatus de Spiritualibus Creaturis,* ed. Leo Keeler. Textus et Documenta, Series Philosophica, No. 13. Rome: Apud Aedes Universitatis Gregorianae, 1946.

— *Super Librum de Causis Expositio,* ed. H. D. Saffrey. Fribourg: Imprimerie St-Paul, 1954.

— *Super Evangelium S. Joannis Lectura,* ed. R. Cai, Va revisa. Turin: Marietti, 1952.

— *Basic Writings of St. Thomas Aquinas.* 2 vols. Edited and Annotated, with an Introduction, by Anton C. Pegis. New York: Random House, 1944.

— *On the Truth of Catholic Faith,* Book 2, tr. J. F. Anderson, New York: Doubleday Image Books, 1956.

William of Auvergne. *Opera Omnia.* 2 vols. Aureliae: F. Hotot, 1674.

II. SECONDARY SOURCES

1. Books

Adler, Mortimer. *St. Thomas and the Gentiles.* Milwaukee: Marquette University Press, 1938.

Anawati, M. M. et L. Gardet. *Introduction à la théologie musulmane.* Paris: J. Vrin, 1948.

Anderson, James. *The Cause of Being.* St. Louis: Herder, 1952.

Antweiler, Anton. *Die Anfangslosigkeit der Welt nach Thomas von Aquin und Kant.* Trier: Paulinus, 1961.
Armstrong, A. H. and R. A. Markus. *Christian Faith and Greek Philosophy.* London: Darton, Longman, Todd, 1960.
Barnett, L. *The Universe and Dr. Einstein.* New York: The New American Library, 1954.
Bergmann, Gustav. *The Metaphysics of Logical Positivism.* New York: Longmans, Green and Co., 1954.
Bergson, Henri. *Creative Evolution,* tr. A. Mitchell. New York: Holt, 1911.
Biemel, W. *Die Idee der Phänomenologie.* The Hague: M. Nijhoff, 1950.
Bofill, J. *La escala de los seres, o el dinamismo de la perfeccion.* Barcelona: Ariel, 1950.
Busa, Roberto. *La terminologia tomistica dell'interiorita.* Saggi de metodo per un'-interpretazione della metaphysica della presenza. Milano: Bocca, 1949.
Carlo, William. *The Ultimate Reducibility of Essence to Existence in Existential Metaphysics.* The Hague: M. Nijhoff, 1966.
Carroll, Owen. *History of the Development of Metaphysics in the Commentary on Peter Lombard's Sentences by Thomas Aquinas.* Ottawa: L'Université d'Ottawa (manuscript), 1958.
Chenu, M.-D. *Introduction à l'étude de saint Thomas d'Aquin.* Montréal: Institut d'Etudes Médiévales, 1950.
Chevalier, Jacques. *Histoire de la Pensée.* 3 vols. Paris: Flammarion, 1955-56.
Coreth, Emerich, *Metaphysik. Eine methodisch-systematische Grundlegung.* Innsbruck: Tyrolia-Verlag, 1964.
Del Prado, Norbert. *De Veritate fundamentali philosophiae christianae.* Fribourg: Ex typis consociationis Sancti Pauli, 1911.
Dolch, Heimo. *Kausalitat in Verstandnis des Theologen und der Begrunder neuzeitlichen Physik.* Freiburg-im-Breisgau: Herder, 1954.
Eddington, Arthur S. *The Philosophy of Physical Science.* Ann Arbor: The University of Michigan Press (C.U.P., 1939), 1958.
— *The Expanding Universe.* Ann Arbor: The University of Michigan Press (C.U.P., 1933), 1958.
Fabro, Cornelio. *Participation et Causalité selon S. Thomas.* Louvain: Publications Universitaires, 1961.
Fakhry, Majid. *Islamic Occasionalism and its critique by Averroes and Aquinas.* London: Allen and Unwin, 1958.
Finance, Joseph de. *Etre et Agir dans la philosophie de saint Thomas.* Paris: Beauchesne, 1945.
Forest, Aime. *La structure métaphysique du concret selon saint Thomas d'Aquin.* Paris: Vrin, 1956.
Gardet, Louis and M. M. Anawati. *Introduction à la théologie musulmane.* Paris: J. Vrin, 1948.
Geiger, L.-B. *La participation dans la philosophie de S. Thomas d'Aquin,* Paris: J. Vrin, 1942.
Gierens, M. *Controversia de aeternitate mundi.* Textus et documenta, Series Phil., No. 6. Roma: Pontificia Universitas Gregoriana, 1933.
Gilson, Etienne. *A History of Christian Philosophy in the Middle Ages.* New York: Random House, 1955.
— *Being and Some Philosophers.* Toronto: Pontifical Institute of Mediaeval Studies, 1949.
— *Elements of Christian Philosophy.* New York: Doubleday and Co., 1960.
— *L'être et l'essence.* Paris: J. Vrin, 1962.

Goichon, A.-M. *La distinction de l'essence et de l'existence d'après Ibn Sina (Avicenne).* Paris: Desclée de Brouwer. 1937.
— *La philosophie d'Avicenne, et son influence en Europe mediévale,* 2e ed. rev. Paris: Adrien-Maisonneuve, 1951.
Grabmann, M. *I divieti eccleciastici di Aristotele sotto Innocenzo III e Gregorio IX,* in *Miscellanea hist. pontif.* V, fasc. 1. Rome: Saler, 1941.
Hayen, André. *La Communication de l'être d'après S. Thomas d'Aquin.* 2 Vols. Bruges: Desclée de Brouwer, 1957.
— *L'intentionnel selon saint Thomas.* Bruges: Desclée de Brouwer, 1954.
Henle, Robert. *St. Thomas and Platonism.* The Hague: M. Nijhoff, 1956.
Hoenen, Peter. *Cosmologia.* Rome: Aedes Pont. Universitatis Gregorianae, 1956.
Hoyle, Fred. *The Nature of the Universe.* Oxford: Blackwell, 1951.
Husik, Isaac. *A History of Mediaeval Jewish Philosophy.* New York: Meridian Books, 1958.
Jalbert, Guy. *Nécessité et Contingence chez Saint Thomas d'Aquin et chez ses Prédécesseurs.* Ottawa: Editions de l'Université, 1961.
Jaworski, M. *Arystotelesowska I Tomistyczna. Teoria Pryzyczny Sprawczej na Tle Pojecia Bytu.* Lublin: Towarzystwo Naukowe Katolickiego Universytetu Lubelskiego, 1958.
Jeans, J. *The Universe Around Us.* New York: The Macmillan Co., 1947.
Jolivet, Regis. *Essai sur les rapports entre la pensée grecque et la pensée chrétienne.* Paris: J. Vrin, 1955.
— *La notion de substance.* Paris: Beauchesne, 1929.
Krempel, A. *La doctrine de la relation chez saint Thomas; exposé historique et systématique.* Paris: J. Vrin, 1952.
Krenn, Kurt. *Vermittlung und Differenz?* Von Sinn des Seins in der Befundlichkeit der Partizipation beim hl. Thomas von Aquin. Roma: Pontificia Universitas Gregoriana, 1962.
Kwant, Remy. *De Gradibus Entis.* Amsterdam: H. J. Paris, 1946.
Labriolle, Pierre de. *La réaction païenne, Etude sur la polémique anti-chrétienne du Ier au Ve siècle.* Paris: L'artisan du livre, 1934.
Lee, Otis. *Existence and Inquiry; a study of thought in the modern world.* Chicago: University of Chicago Press, 1949.
Legrand, Joseph. *L'univers et l'homme dans la philosophie de saint Thomas.* Bruxelles: Desclée de Brouwer, 1946.
Litt, Thomas. *Les Corps Célestes dans l'Univers de Saint Thomas d'Aquin.* Louvain: Publications Universitaires, 1963.
Loriaux, R. *L'être et la forme selon Platon, essai sur la dialectique platonicienne.* Paris: Desclée de Brouwer, 1955.
Lyttkens, H., *Analogy between God and the World; an investigation of its background and interpretation of its use by Thomas of Aquino.* Uppsala: Almqvist, 1952.
Madkour, I. *La place d'Alfarabi dans l'école philosophique musulmane.* Paris: Adrien-Maisonneuve, 1934.
Mainberger, G. *Die Seinsstufung als Methode und Metaphysik.* Freiburg (Schweiz): Universitäts-verlag, 1959.
Marcel, Gabriel. *Le mystère de l'être.* Paris: Aubier, 1951.
— *L'homme problématique.* Paris: Aubier, 1955.
Maritain, Jacques. *Quatre essais sur l'esprit dans sa condition charnelle.* Paris: Alsatia, 1956.
Montagnes, Benoit. *La doctrine de l'analogie de l'être d'après saint Thomas d'Aquin.* Louvain: Publications Universitaires, 1963.
Munitz, Milton. *The Mystery of Existence: An Essay in Philosophical Cosmology.* New York: Appleton-Century-Crofts, 1965.

O'Brien, M. C. *The Antecedents of Being, an Analysis of the Concept "de nihilo" in the Philosophy of St. Thomas Aquinas. Washington:* Catholic University Press, 1939.

Owens, Joseph. *St. Thomas and the Future of Metaphysics.* Milwaukee: Marquette University Press, 1957.

Paulus, J. *Henri de Gand, essai sur les tendences de sa métaphysique.* Paris: J. Vrin, 1938.

Peach, J. V. *Cosmology and Christianity.* New York: Hawthorn Books, Inc., 1965.

Pieper, Josef. *The Silence of Saint Thomas,* tr. J. Murray and D. O'Connor. Chicago: Henry Regnery, 1965.

Prezioso, F. *De Aristotelis creationismo secundum S. Bonaventuram et secundum S. Thomam.* Rome: Catholic Book Agency, 1942.

Rescher, Nicholas. *Al-Farabi, An Annotated Bibliography.* Pittsburgh: The University of Pittsburgh Press, 1962.

Riaza, Jose. *El Comienzo del Mundo,* Exposicion a la luz de los avances cientificos. Madrid: Biblioteca de Autores Cristianos, 1959.

Roques, Rene. *L'univers dionysien; structure hiérarchique du monde selon le pseudo-Denys.* Paris: Aubier, 1954.

Russell, Bertrand. *A History of Western Philosophy.* New York: Simon and Schuster, 1945.

Saliba, D. *Etude sur la métaphysique d'Avicenne.* Paris: Presses Universitaires, 1926.

Sertillanges, Antonin D. *L'idée de création et ses retentissements en philosophie.* Paris: Aubier, 1945.

Simonin, H.-D. and G. Meersseman. *De Sacramentorum efficientia apud theologos ordinis praedicatorum,* 1229-1276. Rome: Angelicum, 1936.

Smith, Gerard and Lottie Kendzierski. *The Philosophy of Being.* New York: The Macmillan Co., 1961.

Spiegelberg, Herbert. *The Phenomenological Movement.* The Hague: Martinus Nijhoff, 1960. 2 Vols.

Stallknecht, Newton and Robert Brumbaugh. *The Compass of Philosophy.* New York: Longmans, Green, and Co., 1954.

Tonquedec, Joseph de. *Questions de Cosmologie et de Physique chez Aristote et chez S. Thomas.* Paris: J. Vrin, 1950.

Vajda, Georges. *Introduction à la pensée Juive au moyen age.* Paris: J. Vrin, 1947.

Van Imschoot, P. *La théologie de l'ancien testament.* Tome I, *Dieu.* Tournai: Desclée de Brouwer, 1954.

Van Steenbergen, F. *Aristotle in the West,* tr. L. Johnston. Louvain: Institut Supérieur de Philosophie, 1955.

Vollert, C., L. Kendzierski, P. Byrne. *St. Thomas Aquinas, Siger de Brabant, St. Bonaventure: On the Eternity of the World.* Milwaukee: Marquette University Press, 1964.

Wahl, Jean. *Vers la fin de l'ontologie. Etude sur l'introduction dans la métaphysique par Heidegger.* Paris: Sedes, 1956.

Weizsacker, C. F. von. *The World View of Physics,* tr. M. Grene. Chicago: University of Chicago Press, 1952.

Whittaker, E. *Space and Spirit; theories of the universe and the arguments for the existence of God.* London: Thomas Nelson, 1946.

— *From Euclid to Eddington.* New York: Dover Publications, 1958.

2. *Articles*

Adler, Mortimer. "The Demonstration of God's Existence," *The Thomist,* V (The Maritain Volume, 1943), 188–218.

Albertson, James. "Instrumental Causality in St. Thomas," *New Scholasticism*, XXVIII (1954), 409–35.

Anderson, James. "Creation as a Relation," *New Scholasticism*, XXIV (1950), 263–83.

Ayer, A. J. "The Genesis of Metaphysics," *Philosophy and Analysis*, ed. M. Macdonald. New York: The Philosophical Library, 1954. Pp. 23–25.

Beaurecueil, Serge de. "Gazzali et S. Thomas d'Aquin," *Bulletin francais d'archéologie orientale*, XLVI (1947), 199–238.

Birault, H. "La foi et la pensée d'après Heidegger," *Philosophies chrétiennes* (Recherches et débats, cahier 10). Paris: Arthème Fayard, 1955. Pp. 108–32.

Bode, R. "Creationism in Physics and Philosophy," *Proceedings of the American Catholic Philosophical Association*, XXIX (1955), 133–37.

Bondi, Hermann. "Astronomy and Cosmology," *What Is Science?* ed. James R. Newman. New York: Washington Square Press, 1961. Pp. 69–101.

Braun, E. "Peut-on parler d'existentialisme thomiste? Le problème de l'*esse* chez saint Thomas," *Archives de Philosophie*, XXII (1959), 211–26, 529–65; XXIII (1960), 252–89.

Bremond, André. "La synthèse thomiste de l'acte et de l'idée," *Gregorianum*, XII (1931), 267–83.

Breton, Stanislaus. "L'idée de transcendental," *Recherches de Philosophie* VI. Paris: Desclée de Brouwer, 1963. Pp. 45–74.

Carnap, R. "Uberwindung der Metaphysik durch logische Analyse der Sprache," *Erkenntnis*, II (1932), 219–41.

Catania, Francis. "Albert the Great, Boethius, and Divine Infinity," *Recherches de théologie ancienne et médiévale*, XXVIII (1961), 97–114.

Callus, Daniel. "Les sources de saint Thomas. Etat de la question," *Aristote et saint Thomas d'Aquin*. Louvain: Publications Universitaires, 1957. Pp. 93–174.

Clarke, W. Norris, "The Limitation of Act by Potency: Aristotelianism or Neoplatonism?" in *New Scholasticism*, XXVI (1952), 167–94.

Collins, J. "The Religious Theme in Existentialism," *Philosophy and the Modern Mind*, ed. Francis Canfield. Detroit: Sacred Heart Seminary, 1961. Pp. 20–48.

Corvez, M. "Existence et essence," *Revue thomiste*, LI (1951), 305–330.

Couesnongle, Vincent de. "Mesure et causalité dans la 'quarta via'," *Revue thomiste*, LVIII (1958), 55–74, 244–84.

D'Alverny, M. T. "Anniya-Anitas," *Mélanges offerts à Etienne Gilson*. Paris: J. Vrin, 1959. Pp. 59–91.

Daly, C. B. "New Light on Wittgenstein," *Philosophical Studies* (Maynooth), X (1961), 5–49; XI (1962), 28–62.

Dondaine, H. D. "A propos d'Avicenne et de saint Thomas," *Revue thomiste*, LI (1951), 441–53.

Dondeyne, A. "La difference ontologique chez M. Heidegger," *Revue Philosophique de Louvain*, LVI (1958), 35–62, 251–93.

Dubarle, Dominique. "La causalité dans la philosophie d'Aristote," *Recherches de Philosophie* I. Paris: Desclée de Brouwer, 1955. Pp. 9–55.

Ducoin, Georges. "Saint Thomas commentateur d'Aristote," *Archives de Philosophie*, XX (1957), 78–117, 240–71, 392–445.

Duquesne, M. "A propos de la participation dans la philosophie de saint Thomas" in *Mélanges de science religieuse*, II (1945), 161–79.

Fabro, C. "Un itinéraire de S. Thomas; l'établissement de la distinction réelle entre essence et existence," *Revue de philosophie*, XXXIX (1939), 285–310.

Fakhry, Majid. "The Antinomy of the Eternity of the World in Averroes, Maimonides, and Aquinas," *Museon*, LXVI (1953), 139–55.

Finance, Joseph de. "Remarques sur l'emploi des mots créer et création," *Les Etudes Philosophiques*, XII (1957), 69–72.

Gagnebet, R. "Dieu, sujet de la théologie selon saint Thomas d'Aquin," *Problemi scelti di theologia contemporanea*. Rome: Analecta Gregoriana, Vol. LXVIII, 1954. Pp. 41–55.

Gardet, L. "Quelques aspects de la pensée avicennienne," *Revue thomiste*, XLVII (1939), 539–75, 693–742.

— "En l'honneur du millénaire d'Avicenne," *Revue thomiste*, LI (1951), 333–45.

Gilson, Etienne. "Pourquoi saint Thomas a critiqué saint Augustin," *Archives d'histoire doctrinale et littéraire du moyen âge*, I (1926), 5–127.

— "Maimonide et la Philosophie de l'Exode," *Mediaeval Studies*, XIII (1951), 223–25.

— "Notes pour l'histoire de la cause efficiente," *Archives d'histoire doctrinale et littéraire du moyen âge*, XXIX (1962), 7–31.

— "Les principes et les causes," *Revue thomiste*, LII (1952), 39–63.

— "De la notion d'être divin dans la philosophie de saint Thomas d'Aquin," *Doctor Communis*, XVIII (1965), Vol. I, 113–29.

Harasta, K. "Die Bedeutung Maimuns für Thomas von Aquin," *Judaica* (Zürich), XI (1955), 65–83.

Isaac, J. "L'unité de l'être," *Bulletin thomiste*, VIII)1947-53), 52–55.

Johann, R. "A Comment on Secondary Causality," *The Modern Schoolman*, XXV (1947-48), 19–25.

Journet, Charles. "L'univers de création ou l'univers antérieur à l'église," *Revue thomiste*, LIII (1953), 349–87.

Kendzierski, Lottie H. "Maimonides' Interpretation of the 8th Book of Aristotle's Physics," *New Scholasticism*, XXX (1956), 37–48.

— "The Doctrine of Eternal Matter and Form," *The Modern Schoolman*, XXXI (1953-1954), 171–183.

— "Eternal Matter and Form in Siger of Brabant," *The Modern Schoolman*, XXXII (1954-1955), 223–241.

Klubertanz, George. "Esse-Existere in St. Bonaventure," *Medieval Studies*, VIII (1946), 169–88.

Klubertanz, George. "Where is the Evidence for Thomistic Metaphysics?" *Revue Philosophique de Louvain*, LVI (1958), 294–315.

Kossel, Clifford. "St. Thomas's Theory of the Causes of Relations," *The Modern Schoolman*, XXV (1947-48), 151–72.

Kramer, K. "Die Creatio nach Thomas von Aquin und dem Liber de Causis," *Ekklesia*. Trier: Paulinus-verlag, 1962. Pp. 321–44.

Kwant, R. "The Ambiguity of the Fact," *The Modern Schoolman*, XXXVII (1959-60), 287–300.

Langan, T. "Heidegger in France," *The Modern Schoolman*, XXXIII (1955-56), 114–18.

Le Bachelet, X. "Dieu: sa nature d'après les Pères," *Dictionnaire de théologie catholique*, IV, 1023–1154.

Lonergan, Bernard. "St. Thomas' Theory of Operation," *Theological Studies*, III (1942), 375–402.

Longpré, E. "Thomas d'York et Matthieu d'Aquasparta. Textes inédits sur le problème de la création," *Archives d'histoire doctrinale et littéraire du moyen âge*, I (1926), 269–308.

Lotz, Johann B. "Heidegger und das Sein," *Universitas*, VI (1951), 727–34, 839–45. Tr. and rev. "Heidegger et l'être," *Archives de Philosophie*, XIX (1956), 3–23.

Lynch, Lawrence, "Martin Heidegger: Language and Being," *An Etienne Gilson Tribute*. Milwaukee: Marquette University Press, 1959. Pp. 134–47.

Malevez, L. "Le croyant et le philosophe," *Nouvelle Revue Théologique*, LXXXII

(1960), 897–917. English tr. "The Believer and the Philosopher," *Philosophy Today,* V (1961), 14–30.

Mansion, Auguste. "Le Dieu d'Aristote et le Dieu des chrétiens," *La Philosophie et ses problèmes.* Recueil d'études offert à Monseigneur R. Jolivet. Paris: Emmanuel Vitte, 1960. Pp. 21–44.

Mansion, Suzanne. "Positions maitresses d'Aristote," *Aristote et saint Thomas d'Aquin.* Louvain: Publications Universitaires, 1957. Pp. 43–91.

McKeon, Richard. "Being, Existence and That Which Is," *Review of Metaphysics,* XIII (1960), 539–554.

McWilliams, J. "Action Does Not Change the Agent" *Philosophical Studies in Honor of the Very Reverend Ignatius Smith, O. P.* Westminster, Md.: The Newman Press, 1952. Pp. 208–21.

Menne, A. "The Logical Analysis of Existence," *Logico-Philosophical Studies,* ed. A. Menne. Dordrecht: Reidel Publishing Company, 1962. Pp. 88–96.

Metz, André. "Causalité scientifique et cause première," *Archives de Philosophie,* XXIV (1961), 517–41.

Miller, R. "The Empiricist's Dilemma," *Proceedings of the American Catholic Philosophical Association,* XXIX (1955), 157–76.

Motte, A. R. "Théodicée et théologie chez saint Thomas," *Revue des sciences philosophiques et théologiques,* XXVI (1937), 5–26.

O'Shaughnessy, Thomas. "St. Thomas' changing Estimate of Avicenna's Teaching on Existence as an Accident," *The Modern Schoolman,* XXXVI (1959), 245–60.

Owens, J. "The Intelligibility of Being," *Gregorianum,* XXXVI (1955), 169–93.

Peach, J. V. "The Age of the Universe," *Heythrop Journal,* III (1962), 111–25.

Pegis, Anton. "A note on St. Thomas, Summa Theologica I, 44, 1–2," *Medieval Studies,* VIII (1946), 159–68.

— "St. Thomas and the Origin of the Idea of Creation," *Philisophy and the Modern Mind,* ed. Francis Canfield, Detroit: Sacred Heart Seminary, 1961. Pp. 49–65.

— "The Argument of the Proslogion," *Mediaeval Studies,* XXVIII (1966), 228–67.

Phelan, Gerald. "The Existentialism of St. Thomas," *Proceedings of the American Catholic Philosophical Association,* XXI (1946), 25–40.

— "The Being of Creatures," *Proceedings of the American Catholic Philosophical Association,* XXXI (1957), 118–25.

Philippe, M. D. "La participation d'après Aristote," *Revue thomiste,* XLIX (1949), 254–77.

Raeymaeker, Louis de. "Les causes et la causalité absolue," *Revue Philosophique de Louvain,* LV (1957), 153–70. English tr. "The Metaphysical Problem of Causality," *Philosophy Today,* I (1957), 219–29.

Rohmer, Jean. "Astronomie et philosophie," *Revue des Sciences Religieuses,* XXX (1956), 113–34. English tr. *Philosophy Today,* I (1957), 279–89.

Roland-Gosselin, M.-D. "De distinctione inter essentiam et esse apud Avicennam et S. Thomam," *Xenia Thomistica.* Rome: Collegium Angelicum, 1925. Vol. III, pp. 281–88.

Schmitz, Kenneth. "Weiss and Creation," *The Review of Metaphysics,* XVIII (1964), 147–69.

— "Metaphysical Restoration of Natural Things," *An Etienne Gilson Tribute,* ed. Charles O'Neil. Milwaukee: Marquette University Press, 1959. Pp. 245–62.

Sertillanges, A. "La notion de création," *Revue thomiste,* XXXV (1930), 48–57.

Smith, Gerard. "Avicenna and the Possibles," *New Scholasticism,* XVII (1943), 340–57.

Smith, J. Macdonald. "Can Science Prove that God Exists?" *Heythrop Journal,* III (1962), 126–38.

Sweeney, Leo. "Existence/Essence in Thomas Aquinas' Early Writings," *Proceedings of the American Catholic Philosophical Association*, XXXVII (1963), 97–131.
— "Doctrine of Creation in the *Liber de Causis*," *An Etienne Gilson Tribute*, ed. Charles O'Neil. Milwaukee: Marquette University Press, 1959. Pp. 274–89.
— "Research Difficulties in the *Liber de Causis*," *The Modern Schoolman*, XXXVI (1958-59), 109–16.
— "John Damascene's 'Infinite Sea of Essence'," *Studia Patristica*. Berlin: Akademie Verlag, 1962. Vol. VI, pp. 248–63.
— "The Mystery of Existence," *The Modern Schoolman*, XLIV (1966), 57–73.
Thomas, Ivo. "Existence and Coherence," *Methodos*, II (1950), 76–84.
Trouillard, Jean. "L'intelligibilité proclusienne," *La philosophie et ses problèmes*. Paris: Emmanuel Vitte, 1960. Pp. 83–97.
Vajda, Georges. "Notes d'Avicenne sur la Théologie d'Aristote," *Revue thomiste*, LI (1951), 346–406.
Van Steenberghen, Ferdinand. "Sciences positives et existence de Dieu," *Revue Philosophique de Louvain*, LVII (1959), 397–414.
Verneaux, Roger. "Esquisse d'une ontologie du créé," *Revue des sciences religieuses*, XXIV (1950), 301–14.
Vogel, C. J. de. "La theorie de l' ἄπειρον chez Platon et dans la tradition platonicienne," *Revue Philosophique*, LXXXIV (1959), 21–39.
Wenzl, A. "Einstein's Theory of Relativity Viewed from the Standpoint of Critical Realism and Its Significance for Philosophy," *Albert Einstein, Philosopher-Scientist*, ed. P. A. Schilpp. New York: Harper Torchbooks, 1959, pp. 583–606.
White, Morton. "New Horizons in Philosophy," *Saturday Evening Post*, CCXXXIII (Sept. 17, 1960), 24–25 ff.
Wright, Thomas. "Necessary and Contingent Being in St. Thomas," *New Scholasticism*, XXV (1951), 439–66.
Zedler, Beatrice. "St. Thomas and Avicenna in the 'De Potentia Dei'," *Traditio*, VI (1948), 105–59.

NAME INDEX

SYSTEMATIC INDEX

GPSR Compliance

The European Union's (EU) General Product Safety Regulation (GPSR) is a set of rules that requires consumer products to be safe and our obligations to ensure this.

If you have any concerns about our products, you can contact us on ProductSafety@springernature.com

In case Publisher is established outside the EU, the EU authorized representative is:

Springer Nature Customer Service Center GmbH
Europaplatz 3
69115 Heidelberg, Germany

Batch number: 10620141

Printed by Printforce, the Netherlands